Lecture Notes in Control and Information Sciences

Edited by M. Thoma and A. Wyner

Vol. 62: Analysis and Optimization
of Systems
Proceedings of the Sixth International
Conference on Analysis and Optimization
of Systems
Nice, June 19–22, 1984
Edited by A. Bensoussan, J. L. Lions
XIX, 591 pages. 1984.

Vol. 63: Analysis and Optimization
of Systems
Proceedings of the Sixth International
Conference on Analysis and Optimization
of Systems
Nice, June 19–22, 1984
Edited by A. Bensoussan, J. L. Lions
XIX, 700 pages. 1984.

Vol. 64: Arunabha Bagchi
Stackelberg Differential Games
in Economic Models
VIII, 203 pages, 1984

Vol. 65: Yaakov Yavin
Numerical Studies
in Nonlinear Filtering
VIII, 273 pages, 1985.

Vol. 66: Systems and Optimization
Proceedings of the Twente Workshop
Enschede, The Netherlands, April 16–18, 1984
Edited by A. Bagchi, H. Th. Jongen
X, 206 pages, 1985.

Vol. 67: Real Time Control of Large Scale Systems
Proceedings of the First European Workshop
University of Patras, Greece, Juli 9–12, 1984
Edited by G. Schmidt, M. Singh, A. Titli,
S. Tzafestas
XI, 650 pages, 1985.

Vol. 68: T. Kaczorek
Two-Dimensional Linear Systems
IX, 397 pages, 1985.

Vol. 69: Stochastic Differential Systems –
Filtering and Control
Proceedings of the IFIP-WG 7/1 Working Conference
Marseille-Luminy, France, March 12-17, 1984
Edited by M. Metivier, E. Pardoux
X, 310 pages, 1985.

Vol. 70: Uncertainty and Control
Proceedings of a DFVLR International Colloquium
Bonn, Germany, March, 1985
Edited by J. Ackermann
IV, 236 pages, 1985.

Vol. 71: N. Baba
New Topics in Learning Automata
Theory and Applications
VII, 231 pages, 1985.

Vol. 72: A. Isidori
Nonlinear Control Systems:
An Introduction
VI, 297 pages, 1985.

Vol. 73: J. Zarzycki
Nonlinear Prediction
Ladder-Filters for Higher-Order
Stochastic Sequences
V, 132 pages, 1985.

Vol. 74: K. Ichikawa
Control System Design based on
Exact Model Matching Techniques
VII, 129 pages, 1985.

Vol. 75: Distributed Parameter
Systems
Proceedings of the 2nd International
Conference, Vorau, Austria 1984
Edited by F. Kappel, K. Kunisch,
W. Schappacher
VIII, 460 pages, 1985.

Vol. 76: Stochastic Programming
Edited by F. Archetti, G. Di Pillo,
M. Lucertini
V, 285 pages, 1986.

Vol. 77: Detection of
Abrupt Changes in Signals
and Dynamical Systems
Edited by M. Basseville,
A. Benveniste
X, 373 pages, 1986.

Vol. 78: Stochastic
Differential Systems
Proceedings of the 3rd Bad Honnef
Conference, June 3–7, 1985
Edited by N. Christopeit, K. Helmes,
M. Kohlmann
V, 372 pages, 1986.

Vol. 79: Signal
Processing for Control
Edited by K. Godfrey, P. Jones
XVIII, 413 pages, 1986.

Vol. 80: Artificial Intelligence
and Man-Machine Systems
Edited by H. Winter
IV, 211 pages, 1986.

For information about Vols. 1–61 please contact your bookseller or Springer-Verlag.

Lecture Notes in Control and Information Sciences

Edited by M. Thoma and A. Wyner

119

T. S. Başar, P. Bernhard (Eds.)

Differential Games and Applications

Springer-Verlag
Berlin Heidelberg GmbH

Editors

Tamer S. Başar
Coordinated Science Laboratory
University of Illinois at Urbana-Champaign
1101 West Springfield Avenue
Urbana, IL 61801
USA

Pierre Bernhard
INRIA – Sophia Antipolis
2004, Route des Lucioles
06565 Valbonne Cedex
France

ISBN 978-3-540-50758-1

Library of Congress Cataloging in Publication Data

Differential games and applications / T. S. Başar, P. Bernhard (eds.)
(Lecture notes in control and information sciences ; 119)
"All but one of these [articles] are based on presentations made
at the Third International Symposium on Differential Games and
Applications held at INRIA, Sophia-Antipolis, France, on June 16–17,
1988 . . . [and] sponsored by INRIA"–Pref.
ISBN 978-3-540-50758-1 ISBN 978-3-540-46079-4 (eBook)
DOI 10.1007/978-3-540-46079-4

1. Differential games – Congresses. I. Başar, Tamer.
II. Bernhard, P. (Pierre)
III. International Symposium on Differential Games and Applications (3rd : 1988 : INRIA)
IV. Institut national de recherche en informatique et en automatique (France) V. Series.
QA272.D53 1989
519.3–dc19 88-35985

Originally published by Springer-Verlag Berlin Heidelberg New York in 1989

2161/3020-543210

PREFACE

This volume contains fifteen articles on the topic of *differential and dynamic games*, focusing on both theory and applications. All but one of these are based on presentations made at the *Third International Symposium on Differential Games and Applications* held at *INRIA*, Sophia-Antipolis, France, on June 16-17, 1988. The meeting was sponsored by *INRIA*, and involved the participation of *IFAC* and *the IEEE Control Systems Society*.

The first two articles in the volume deal with discrete two-person constant-sum dynamic games, that is games where the action spaces of the players are finite and decisions are made at discrete instants of time. The first paper, which comprises Chapter 1, describes a parlor game named after Emanuel Lasker; it discusses the challenges involved in obtaining a solution for the general case, and computes the value function for a special case of the problem. The second chapter deals with a 'search game' with delayed information for the 'searcher', and obtains expressions for the saddle-point solution in Markovian-type strategies when the discrete positions are either on the line or on the plane, and the 'searcher' is not allowed to use memory. The paper also discusses the implications of allowing the 'searcher' to use memory strategies, and of introducing different types of 'noise' into the perfect (but delayed) measurements of the 'searcher'.

The next six chapters deal with different aspects of zero-sum differential games of the pursuit-evasion type, the class of games introduced and pioneered by Rufus Isaacs in the 1950's. The first of these (chapter 3) outlines a rigorous approach towards formulating and analyzing such games, which involves a discrete-time approximation to the original continuous-time game. The strategy choices by the two players are allowed to depend on the history of the game, and the solution to the original game is captured by letting the length of the small time intervals on which the sequence of discrete-time games is defined go to zero. The next chapter studies the Isaacs partial differential equation that arises in these games, and provides a characterization of its classical and semicontinuous solutions. In this analysis, the author uses contingent epiderivatives, in place of the classical derivatives, which allows him to characterize the solutions which are indicators of closed subsets (corresponding to some desirable behavior), and are only semicontinuous. Chapter 5 also deals with the Isaacs equation, and shows that the capture time of a pursuit-evasion game is the unique viscosity solution of that equation under some singular boundary conditions, and provided that a 'capturability' condition is satisfied. The topic of Chapter 6 is the well-known 'Lion and Man' pursuit-evasion game, the formulation here giving speed advantage to the Lion (the pursuer) by a factor of 1.2. The author shows that the solution to the game features a 'focal surface' of 'radial' positions leading to curved motion by both players until termination, and an 'indifference zone' for Man. Chapter 7 introduces a new approach for the problem of feedback guidance during the pre-launch maneuvering phase of medium range missile combat, which exploits any possible nonoptimal play by the opponent, and is related to the ideas of threat reciprocity, reprisal strategies and a bi-criteria formulation of pursuit-evasion games. Feedback guidance law determination is also the subject of Chapter 8, which obtains a closed-form solution for a coplanar pursuit-evasion game between a coasting missile and a maneuvering target of constant speed, leading to an optimal strategy for the missile as a function of the current and final values of the state variables. The implementable feedback guidance law constructed from this strategy is shown to perform better than proportional navigation, and only slightly worse than the optimal game strategy.

The next two chapters are devoted to stochastic games. The class of stochastic differential games treated in Chapter 9 are piecewise deterministic, with the stochasticity being due to random jump process which determines the times and the nature of switch from one system (or game) to another. In this context, the author studies the existence and characterization of Nash equilibria under feedback and piecewise open-loop strategies. Chapter 10, on the other hand, deals with discrete discounted-cost zero-sum stochastic games (finite state and action spaces), and presents numerical results on the computation of stationary equilibrium strategies, using a recently proposed modified Newton method.

Chapter 11 deals with an application of the nonzero-sum discrete-time dynamic game theory on a model arising in the European gas market, and shows that under some conditions the open-loop and feedback Nash equilibria coincide. Chapter 12 involves a dynamic game formulation where the players are allowed to switch between cooperative and noncooperative (Nash) modes of play, with cooperation occurring if all players benefit from it (as compared with the Nash solution). In the context of linear-quadratic games, it is shown that the evolution of the game depends very much on the type of Pareto-optimal solution adopted for the cooperative mode of play. The next chapter is on an application of differential game theory in predator-prey systems. Using a Lotka-Volterra formulation, the author studies the scenario where the two players, while seeking conflicting growth objectives, cooperate in designing a self-controlled growth policy in order to reach and maintain a common population size level. Chapter 14 presents an extension of the Nash bargaining scheme to discrete-time and continuous-time dynamic games, which uses a two-level decision structure. At the lower level, a weighted sum of the players' objectives is maximized, while at the upper level the bargaining contract is determined, that satisfies the required rationality axioms.

One topic which is considered to belong to the general area of differential games, and which is not treated in the chapters above, is 'stochastic teams', and especially those with decentralized information patterns. The last chapter of this volume provides a selective annotated bibliography of the literature on this topic, covering the past twenty-five years.

Even though the volume is not exhaustive in the choice of topics within the field of differential games and its applications, it does cover a variety of areas and presents recent developments on topics of current interest. Therefore, it should be useful to researchers in differential and dynamic games, systems and control, operations research and mathematical economics. We thank the authors for their contributions to this volume, and to all participants at the *Third International Symposium on Differential Games and Applications* for their active involvement in the discussions during and after the paper presentations.

Our thanks also go to *INRIA*, for both the financial and moral support without which the Differential Games Meeting where these papers were presented could not have been organized. Last, but not least, our sincere thanks go to Ms. Catherine Juncker and Ms. Jacqueline Tchobanian for their invaluable assistance during both the organizational and the execution phases of the Symposium.

August 1988

Urbana, Illinois, USA — Tamer Başar

Sophia Antipolis, Valbonne, France — Pierre Bernhard

LIST OF CONTRIBUTING AUTHORS

AUBIN, J. P.
CEREMADE, Universite de Paris-Dauphine
Paris — France

BANSAL, R.
Coordinated Science Laboratory, University of Illinois
Urbana, Illinois 61801 — USA

BARDI, M.
Dipartimento di Matematica Pura e Applicata, Universita di Padova
1-35131 Padova — Italy

BASAR, T.
Coordinated Science Laboratory, University of Illinois
Urbana, Illinois 61801 — USA

BERGER, E.
DFVLR Institute for Flight Systems Dynamics, Oberpfaffenhofen
8031 Wessling — West Germany

BERKOVITZ, L. D.
Department of Mathematics, Purdue University
West Lafayette, Indiana 47907 — USA

BOJADZIEV, G.
Department of Mathematics and Statistics, Simon Fraser University
Burnaby, British Columbia V5A 1S6 — Canada

BREAKWELL, J. V.
Department of Aeronautics and Astronautics, Stanford University
Stanford, California 94305 — USA

EHTAMO, H.
Systems Analysis Lab, Helsinki University of Technology
02150 Espoo — Finland

FLÅM, S. D.
Institute of Economics, University of Bergen
Bergen — Norway

GAO, L.
Department of Technical Mathematics and Informatics
Delft University of Technology
2600 AJ Delft — The Netherlands

GREEN, A.
Faculty of Aeronautical Engineering, Technion-Israel Institute of Technology
Haifa — Israel

GRIMM, W.
DFVLR Institute for Flight Systems Dynamics, Oberpfaffenhofen
8031 Wessling — West Germany

GUELMAN, M.
Guidance Division, RAFAEL, M.O.D.
Haifa — Israel

HÄMÄLÄINEN
Systems Analysis Lab, Helsinki University of Technology
02150 Espoo Finland

HAURIE, A.
GERAD, Ecole des Hautes Etudes Commerciales
Montreal, Quebec H3T 1V6 Canada

JAKUBOWSKI, A.
Department of Technical Mathematics and Informatics
Delft University of Technology
2600 AJ Delft The Netherlands

KAHN, J.
Rutgers University
New Brunswick, New Jersey 08903 USA

KLOMPSTRA, M. B.
Department of Technical Mathematics and Informatics
Delft University of Technology
2600 AJ Delft The Netherlands

LARGARIAS, J. C.
AT&T Bell Laboratories
Murray Hill, New Jersey 07974 USA

OLSDER, G. J.
Department of Technical Mathematics and Informatics
Delft University of Technology
2600 AJ Delft The Netherlands

PRASAD, U. R.
DFVLR Institute for Flight Systems Dynamics, Oberpfaffenhofen
8031 Wessling West Germany

RUUSUNEN, J.
Systems Analysis Lab, Helsinki University of Technology
02150 Espoo Finland

SHINAR, J.
Faculty of Aeronautical Engineering, Technion-Israel Institute of Technology
Haifa Israel

SORAVIA, P.
Dipartimento di Matematica Pura e Applicata, Universita di Padova
1-35131 Padova Italy

TOLWINSKI, B.
Department of Mathematics, Colorado School of Mines
Golden, Colorado 80401 USA

WITSENHAUSEN, H. S.
AT&T Bell Laboratories
Murray Hill, New Jersey 07974 USA

ZACCOUR, G.
GERAD, Ecole des H.E.C.
Montreal, Quebec H3T 1V6 Canada

TABLE OF CONTENTS

CHAPTERS

On Lasker's Card Game

J. Kahn

Rutgers University
New Brunswick, New Jersey

J. C. Lagarias & H. S. Witsenhausen

AT&T Bell Laboratories
Murray Hill, New Jersey 07974

Abstract

In 1929 Emanuel Lasker, the mathematician and world chess champion, described a two person constant sum perfect information game, which he called *whistette*. The game uses a deck of cards that consists of a single totally ordered suit of $2n$ cards. To begin play the deck is divided into two hands A and B of n cards each, held by players Left and Right, and one player is designated as having the *lead*. The player on lead chooses one of his cards, and the other player after seeing this card selects one of his own to play. The player with the higher card wins a "trick" and obtains the lead. The cards in the trick are removed from each hand, and play then continues until all cards are exhausted. Each player strives to maximize (or minimize, in the misère version) his trick total, and the *value* of the game to each player is the number of tricks he takes under optimal play. While we have found an optimal strategy for the misère version, the regular game remains unsolved in general. In previous papers we derived basic properties of the regular game, found criteria under which one hand is guaranteed to be better than another, and determined the value functions in several special cases. In this paper we compute the value function for the "2 blocks vs 3" case.

1. Introduction

We consider the following two-person constant sum perfect information game, which was described in 1929 by the mathematician and world chess champion Emanuel Lasker [4], who called it *whistette*, and studied under the name *End Play Game* in [1]-[3]. The players are called L (left) and R (right). In the starting position each player holds n cards (his "hand"). The $2n$ cards are totally ordered. Both players know the contents of both hands and the order. One of the players is given the initial lead. Play proceeds as in whist, bridge, etc. The leader selects one of his cards, and the other player, *after* he has seen this selected card, plays one of his cards in response. The highest of the two cards scores a "trick" for the player who played it and the lead passes to this player. The cards played are removed, producing a new position with hands of $n-1$ cards and a specified player on lead, from which play proceeds until the cards are exhausted. The players' objective is to score as many tricks as possible.

To further study the game, let (ε, A, B) denote a *state* of the game, in which ε indicates who has the lead (ε=0 if R leads, ε=1 if L leads), A the n cards held by L and B the n cards held by R, and let $V_n^{\varepsilon}(A,B)$ denote the *value* of such a state to Left, i.e. the number of tricks Left takes assuming optimal play by both players. The position that appears to Left as (ε, A, B), when viewed from Right's side, appears as $(1-\varepsilon, B, A)$. Therefore, one has the identity

$$V_n^{\varepsilon}(A, B) + V_n^{1-\varepsilon}(B, A) = n. \tag{1.1}$$

The basic problems for this game are to find an algorithm that quickly computes the value of the game for each player, and, if possible, to find a simple optimal strategy for each player. Neither of these problems has been solved, and [1] presents evidence that both problems are computationally difficult.

One approach to studying the value function is to analyze hands having a simple structure. A sequence of consecutive cards in the same hand will be called a *block*, and hands may be ranked in complexity by the number of blocks they contain. If one player has k blocks the other has at most $k+1$ blocks. For a single block hand, the value function is trivial and lead-independent, whether the other hand has two blocks or one. Consider the case of two blocks in each hand as the first non-trivial one. Call the player with the lowest block Left. Thus Left has a block of $n-a$ lowest cards and a block of a higher cards between which lies a block of b cards for Right, who also holds the $n-b$ highest cards. Thus

$$A = \{n+b, n+b-1, \ldots, n+b-a+1\} \cup \{n-a, n-a-1, \ldots, 1\}$$

$$B = \{2n, 2n-1, \ldots, n+b+1\} \cup \{n-a+b, \ldots, n-a+1\} \tag{1.2}$$

For such A, B define the function

$$B_n^{\varepsilon}(a, b) = V_n^{\varepsilon}(A, B)\,. \tag{1.3}$$

In [1] we calculated the value of the game for two blocks in each hand.

Theorem 1. One has

$$B_n^{\varepsilon}(a, b) = \max\{a+b-n,\ \min(a, b, [\frac{a+b+1-\varepsilon}{3}])\}\,, \tag{1.4}$$

where $[x]$ *denotes the integer part of* x.

In the rest of this paper we obtain the value of the game for two blocks in one hand and three in the other. (The result was stated without proof in [1].) The answer (Theorem 2) has a rather complicated form which suggests that no simple formula for the value exists. The theorem is used in [2] to give a simple proof of the result that two partial orderings, called *superiority* and *precedence*, are equivalent.

2. Solution for Two Blocks vs. Three

Assume Left's hand consists of two blocks, the higher of which has a cards. These blocks separate Right's hand into three blocks, one of b lowest cards, one of c cards between Left's blocks, and one of $n-b-c$ highest cards. That is, for

$$\begin{cases} 0 \le a \le n \\ 0 \le b \\ 0 \le c \\ b+c \le n \end{cases} \tag{2.1}$$

the hands are

$$A = \{n+b+c,\ n+b+c-1, \ldots, n+b+c-a+1\} \cup \{n+b-a,\ n+b-a-1, \ldots, b+1\}$$

and

$$B = \{2n, 2n-1, \ldots, n+b+c+1\} \cup \{n+b+c-a, \ldots, n+b-a+1\} \cup \{b, b-1, \ldots, 1\}. \tag{2.2}$$

For such A, B define

$$C_n^\varepsilon(a, b, c) = V_n^\varepsilon(A, B)\ . \tag{2.3}$$

We shall use the notation

$$\delta_k = 1 \text{ for } k = 0\ ,\quad 0 \text{ otherwise}\ . \tag{2.4}$$

The value for the game with two blocks versus three is as follows.

Theorem 2. One has, for $n > 0$, $\varepsilon \in \{0,1\}$ and a,b,c as in (2.1):

$$C_n^\varepsilon(a, b, c) = \max\{b,\ a+b+c-n,\ \min(b+c,\ a+b,\ [\frac{a+2b+c+1-\varepsilon+\delta_{n-b-c}}{3}])\} \tag{2.5}$$

Note that when $\max(a, c) < n - b$, the value (2.5) is invariant under interchange of a and c.

Proof:

First verify that (2.5) is correct in the boundary cases where at least one block is null. There are five such cases in which we must have

$$C_n^\varepsilon(0, b, c) = b \tag{2.6}$$

$$C_n^\varepsilon(n, b, c) = b+c \tag{2.7}$$

$$C_n^\varepsilon(a, b, 0) = b \tag{2.8}$$

$$C_n^\varepsilon(a, 0, c) = B_n^\varepsilon(a, c) \tag{2.9}$$

$$C_n^\varepsilon(a, n-c, c) = n - B_n^{1-\varepsilon}(c, n-a) \tag{2.10}$$

where the right side of (2.10) is derived from Theorem 1 via (1.1). Now (2.6), (2.7), (2.8) hold by inspection. For (2.9) the only flaw is the presence of δ_{n-c} where we should have 0. But this matters only when $c = n$ in addition to $b = 0$ and then (2.5) reduces to a, the correct value.

The left side of (2.10) is

$$\max(n-c,\ a,\ \min(n,\ a+n-c,\ [\frac{a+2n-c+2-\varepsilon}{3}])) \tag{2.11}$$

and the right side is

$$n - \max(c-a,\ \min(c,\ n-a,\ [\frac{c+n-a+\varepsilon}{3}]))$$

$$= \min(n+a-c,\ \max(n-c,\ a,\ n - [\frac{c+n-a+\varepsilon}{3}])) \tag{2.12}$$

Now

$$[\frac{a+2n-c+2-\varepsilon}{3}] = n - [\frac{c+n-a+\varepsilon}{3}] \tag{2.13}$$

because if $c+n-a = 3m+r$, $0 \le r \le 2$ then letting $r+\varepsilon = x$, (2.13) reduces to

$$[\frac{2-x}{3}] = -[\frac{x}{3}] \tag{2.14}$$

which holds for all integer x.

Denoting the left side of (2.13) by μ we have $\mu \le n$ by (2.13) itself. Using this and the distributivity of the lattice of reals under (min,max), (2.11) becomes

$$\min(\max(n-c,\ a,\ a+n-c),\ \max(n-c,\ a,\ \mu))$$

$$= \min(a+n-c,\ \max(n-c,\ a,\ \mu)),$$

which by (2.13) is (2.12), proving (2.10).

The theorem holds for $n < 3$ because all cases are then boundary cases. Suppose it holds up to $n-1$. As the boundary has been verified for all n, one may assume for n, that

$$\begin{aligned} 1 &\le a \le n-1 \\ 1 &\le b \\ 1 &\le c \\ b+c &\le n-1 \end{aligned} \tag{2.15}$$

By monotonicity and distributivity, one has

$$C_n^1(a, b, c) = \max(\min(x, y), \min(u, z)) \tag{2.16}$$

and

$$C_n^0(a, b, c) = \min(x, w, \max(y, v)) \tag{2.17}$$

where

$$\begin{aligned} x &= 1 + C_{n-1}^1(a, b-1, c)\ , \\ y &= C_{n-1}^0(a, b, c-1)\ , \\ z &= 1 + C_{n-1}^1(a-1, b-1, c)\ , \\ u &= C_{n-1}^0(a-1, b, c)\ , \\ v &= 1 + C_{n-1}^1(a-1, b, c-1)\ , \\ w &= C_{n-1}^0(a, b, c)\ . \end{aligned} \tag{2.18}$$

We will show now that, subject to (2.15), one has $x \ge y$, $u \ge z$, $v \ge y$ and $w \ge x$. This means that, as long as none of the 5 blocks is exhausted, it is optimal:

(1) for R to take when L leads low,

(2) for R to duck when L leads high,

(3) for L to take when R leads from his middle block,

(4) for R not to lead from his top block.

Indeed, one has

$$x = 1 + \max(b-1,\, a+b+c-n,\, \min(b+c-1,\, a+b-1,\, [\frac{a+2b+c-2+\delta_{n-b-c}}{3}]))$$

$$= \max(b,\, a+b+c-n+1,\, \min(b+c,\, a+b,\, [\frac{a+2b+c+1+\delta_{n-b-c}}{3}])) \,. \tag{2.19}$$

$$y = \max(b,\, a+b+c-n,\, \min(b+c-1,\, a+b,\, [\frac{a+2b+c+\delta_{n-b-c}}{3}])) \,. \tag{2.20}$$

Termwise comparison of (2.19) and (2.20) shows that

$$x \geq y \,. \tag{2.21}$$

One has

$$z = 1 + \max(b-1,\, a+b+c-n-1,\, \min(b+c-1,\, a+b-2,\, [\frac{a+2b+c-3+\delta_{n-b-c}}{3}]))$$

$$= \max(b,\, a+b+c-n,\, \min(b+c,\, a+b-1,\, [\frac{a+2b+c+\delta_{n-b-c}}{3}])) \tag{2.22}$$

while

$$u = \max(b,\, a+b+c-n,\, \min(b+c,\, a+b-1,\, [\frac{a+2b+c+\delta_{n-b-c-1}}{3}])) \tag{2.23}$$

By (2.15) $\delta_{n-b-c} = 0$ while $\delta_{n-b-c-1}$ may be 1. Thus $u = z$ when $b+c < n-1$ and in any case

$$u \geq z \,. \tag{2.24}$$

One has

$$v = 1 + \max(b,\, a+b+c-n-1,\, \min(b+c-1,\, a+b-1,\, [\frac{a+2b+c-2+\delta_{n-b-c}}{3}]))$$

$$= \max(b+1,\, a+b+c-n,\, \min(b+c,\, a+b,\, [\frac{a+2b+c+1+\delta_{n-b-c}}{3}])) \tag{2.25}$$

comparing (2.25) with (2.20) termwise shows that

$$v \geq y \,. \tag{2.26}$$

Also,

$$w = \max(b,\, a+b+c-n+1,\, \min(b+c,\, a+b,\, [\frac{a+2b+c+1+\delta_{n-b-c-1}}{3}])) \tag{2.27}$$

and comparing this with (2.19) we see that for $b+c < n-1$ one has $w = x$ and in any case

$$w \geq x\ . \tag{2.28}$$

By (2.21) and (2.24), (2.16) reduces to

$$C_n^1(a, b, c) = \max(y, z)$$
$$= \max(b, a+b+c-n, \min(b+c-1, a+b, \mu), \min(b+c, a+b-1, \mu)) \tag{2.29}$$

where

$$\mu = [\frac{a+2b+c+\delta_{n-b-c}}{3}]\ . \tag{2.30}$$

By distributivity

$$C_n^1(a, b, c) = \max(b, a+b+c-n, \min(\mu, \max(\min(b+c-1, a+b), \min(b+c, a+b-1)))) \tag{2.31}$$

Now note that, for integer s, t

$$\max(\min(s-1, t), \min(s, t-1)) = \min(s, t) - \delta_{s-t}\ , \tag{2.32}$$

So that when $a \neq c$ (2.31) turns into (2.5) with $\varepsilon = 1$. When $a = c$, the result still holds because then

$$\min(\mu, a+b-1) = \mu\ . \tag{2.33}$$

Indeed, this amounts to

$$[\frac{2(a+b)}{3}] \leq a+b-1 \tag{2.34}$$

which reduces to

$$[1 - \frac{a+b}{3}] \leq 0 \tag{2.35}$$

and this holds because $a+b>1$ by (2.15). This completes the induction for $\varepsilon = 1$.

As for $\varepsilon = 0$, (2.17) reduces by (2.26) and (2.28), to

$$C_n^0(a, b, c) = \min(x, v)\ . \tag{2.36}$$

Let

$$m = \min(b+c, a+b, [\frac{a+2b+c+1+\delta_{n-b-c}}{3}])\ , \tag{2.37}$$

then

$$C_n^0(a, b, c) = \min(\max(m, b, a+b+c-n+1), \max(m, b+1, a+b+c-n))$$
$$= \max(m, \min(\max(b, a+b+c-n+1), \max(b+1, a+b+c-n))) \quad (2.38)$$

The dual of (2.32) is

$$\min(\max(s, t+1), \max(s+1, t)) = \max(s, t) + \delta_{s-t} . \quad (2.39)$$

Thus when $a+c \neq n$ (2.38) reduces to (2.5) with $\varepsilon = 0$. When $a+c = n$ (2.38) reduces to $\max(m, b+1)$ instead of the required $\max(m, b)$. However in that case $m \geq b+1$, i.e.

$$[\frac{a+2b+c+1}{3}] \geq b+1 \quad (2.40)$$

for this reduces to

$$[\frac{n-b-2}{3}] \geq 0 \quad (2.41)$$

which holds because (2.15) yields $b \leq n-1-c \leq n-2$.

This completes the induction, proving Theorem 2. □

References

[1] J. Kahn, J. C. Lagarias and H. S. Witsenhausen, Single-Suit Two-Person Card Play, *Intern. J. of Game Theory*, Vol. 16, pp.291-320, 1987

[2] J. Kahn, J. C. Lagarias and H. S. Witsenhausen, Single Suit Two-Person Card Play II: Domination, *Order*, (to appear)

[3] J. Kahn, J. C. Lagarias and H. S. Witsenhausen, Single Suit Two-Person Card Play III: The Misère Game, (preprint).

[4] E. Lasker, *Encyclopedia of Games, vol. I, Card Strategy*, E.P. Dutton & Co., New York 1929.

ON A CLASS OF ZERO-SUM DISCRETE GAMES WITH DELAYED INFORMATION[†]

Tamer Başar

Decision and Control Laboratory
Coordinated Science Laboratory and the
Department of Electrical and Computer Engineering
University of Illinois
1101 W. Springfield Avenue
Urbana, Illinois 61801 / USA

Abstract

In this paper we obtain the complete solution to the following discrete zero-sum dynamic game with delayed information.

Let $\mathbf{Z}$ be the set of all integers, corresponding to possible (discrete) positions of *Player 1* on the real line. At time t, *Player 1* occupies a position $x = x_t$, and makes a move to a possibly different position $x_{t+1} = x_t + u_t$, where $u_t \in \mathbf{Z}$, such that $|u_t| \leq m$, m being a fixed positive integer. Let n and S also be two positive integers, with $S \leq T$, T being another fixed integer, known to both parties. *Player 2* operates in dark, but he has a flash light which he can use a total of S times to observe the current positions of *Player 1*. At each such instant, and based only on the current position information and without using any memory, *Player 2* attempts to predict his opponent's position n time units later. Every time he predicts correctly, he scores 1 point (otherwise, he receives 0 points), and he has altogether S chances to do so. The objective function is the total number of points scored by *Player 2* in S trials. This is to be minimized by *Player 1* (by a proper choice of his action variables $(u_1, ..., u_{T+n}) =: \mathbf{u}$), and maximized by *Player 2* (by choosing the times of prediction (and observation) as well as the forecasts of positions themselves).

We show that the ***saddle-point value*** of the game (in mixed strategies) is equal to $S/(2mn+1)$, and the saddle point is achieved by Markov chain type strategies for *Player 1*. For $n \geq 2$ the game is not a ***feedback game*** because of the time delay in the acquisition of dynamic information by *Player 2*, and hence it does not allow for a temporal decomposition for the construction of the saddle-point strategies. Accordingly, we develop a new approach, which could be called "overlapping decomposition", that leads to an explicit construction of the equilibrium strategies as well as to the value of the game. The paper also discusses extensions to the cases when *(i) Player 1* moves on the plane, *(ii) Player 2* makes only imperfect (noisy) measurements of *Player 1*'s current positions, and *(iii) Player 2* observes the current positions of *Player 1* at all points in time and makes his time-prediction and position-forecast decisions based on this extended information, using memory. It is shown that in this latter case *Player 2* can generally improve upon his performance by adopting a mixed strategy for the times of prediction.

---

† *This work was performed partially while the author was spending a sabbatical year at INRIA, Sophia Antipolis, France, and it was also partially supported by the Air Force Office of Scientific Research under Grant No. AFOSR 088-0178, through the University of Illinois.*

1. Introduction and Problem Formulation

Consider first the following formulation of a zero-sum discrete dynamic game with two-step delayed information for one of the players. Even though this is not the most general class we will be dealing with in this paper, it captures the essence of our more general formulations in subsequent sections, and provides a convenient setting for the presentation of our main ideas.

Let T be a given positive integer, and $\mathbf{u} := \{u_t\}_{t=1}^{T+2}$, $\mathbf{v} := \{v_t\}_{t=1}^{T}$ be two finite sequences under the control of *Player 1* (minimizer) and *Player 2* (maximizer), respectively. For each t, u_t takes values in the set $U := \{-1, 0, 1\}$, whereas v_t takes values in $V := \mathbf{Z}$, where the latter denotes the set of all integers. Let $\mathbf{x} := \{x_t\}_{t=1}^{T+2}$ be another sequence (to be called the *state sequence*) generated by

$$x_{t+1} = x_t + u_t\,, \quad t = 1, 2, \ldots\,; \quad x_1 = 0 \tag{1}$$

where x_1, the initial state, is initially taken as an unknown quantity. Introduce the function

$$L_t := \begin{cases} 1 & \text{if} \quad v_t = x_{t+2} \\ 0 & \text{otherwise} \end{cases} \tag{2a}$$

for each $t \in \mathbf{T} := \{1, 2, .., T\}$. Let S be a given positive integer not exceeding T (*i.e.* $S \in \mathbf{T}$), and let Θ_S denote any subset of $\mathbf{T}$ with S elements. The interpretation here is that Θ_S is the collection of time instants (and there is S of them) when *Player 2* can observe the state $\{x_t\}$ – this choice being under his control. In view of this, *Player 2* would not have access to the initial state x_1, unless he chooses Θ_S in such a way that $1 \in \Theta_S$. Now define the objective function

$$L := \sum_{t \in \Theta_S} L_t \tag{2b}$$

which will in general depend on the sequences $\mathbf{u}$, $\mathbf{v}$, as well as on the choice of the set Θ_S, for each fixed S. To indicate this dependence explicitly, we write $L = L(\mathbf{u}; \mathbf{v}, \Theta_S)$, where Θ_S will also be chosen by *Player 2*. Note that only those actions of *Player 2* that correspond to the time points belonging to Θ_S affect the value of the objective function, and hence for each realization only S components of $\mathbf{v}$ will play a role in the game.

We allow the sequence $\mathbf{u}$ to depend on the past trajectory generated by $(2a)$, whereas the sequence $\mathbf{v}$ is allowed to depend only on the current value of the state; that is, for mappings $\gamma_t : x^t \mapsto u_t \in U$, $\beta_t : x_t \mapsto v_t \in V$, we have

$$u_t = \gamma_t(x^t)\,, \qquad v_t = \beta_t(x^t)\quad , \tag{3}$$

where $x^t := \{x_1, x_2, .., x_t\}$. Let Γ_t and B_t denote the (finite) sets of all such mappings γ_t and β_t, respectively, and introduce the (product) *policy spaces*

$$\mathbf{\Gamma} := \mathop{\mathrm{X}}_{t=1}^{T+2} \Gamma_t \quad \text{and} \quad \mathbf{B} := \mathop{\mathrm{X}}_{t=1}^{T} B_t$$

wherein respectively $\gamma := \{\gamma_1, \gamma_2, .., \gamma_{T+2}\}$ and $\beta := \{\beta_1, \beta_2, .., \beta_T\}$ lie. The set Θ_S, on the other hand, will be chosen by *Player 2* in an ***open-loop*** fashion at the start of the game. To complete the description, we let $\boldsymbol{\Theta}_S$ denote the (finite) class of all such sets, *i.e.* $\Theta_S \in \boldsymbol{\Theta}_S$, for each fixed $S \in \mathbf{T}$.

The zero-sum game of interest is then the minimization of $L(\gamma(\mathbf{x}); \beta(\mathbf{x}), \Theta_S)$ over $\gamma \in \Gamma$ and the maximization of the same quantity over $(\beta \in \mathbf{B}, \Theta_S \in \boldsymbol{\Theta}_S)$, with $S \in \mathbf{T}$ taken as a fixed integer. Note that even though *Player 2* may not have access to the true value of the initial state, x_1, the objective function is well defined on $(\Gamma, \mathbf{B} \times \boldsymbol{\Theta}_S)$ since the first relevant variable is $x_{\underline{i}}$, where $\underline{i}$ is the minimum element of Θ_S, and for any choice of strategies the value of L is independent of $x_{\underline{i}}$. Now, two observations are in place here. The first is that this is not what is commonly known as a ***perfect state information game***, basically because of the structure of the objective function (2). In fact, letting $\tilde{v}_t := v_{t-2}$, we readily see from the relationship $\tilde{v}_t = \beta_{t-2}(x_{t-2})$ that the problem exhibits an information delay of two time units. The second observation is that the ***upper*** and ***lower values*** of this game (in pure strategies) are different, and hence to obtain a ***saddle point*** we have to allow for the possibility of ***mixed*** strategies.

Let $\mathbf{P}_1$ be the class of all (discrete) probability measures for *Player 1*, defined on the (finite) set Γ, with each element denoted by p_1. Similarly, let $\mathbf{P}_2$ be the class of all (discrete) probability measures for *Player 2*, defined on $(\mathbf{B}, \boldsymbol{\Theta}_S)$, with each such measure denoted by p_2. Note that $\mathbf{P}_2$ actually depends on the integer S, which we do not explicitly indicate here to save from notation. Finally, for each $(p_1, p_2) \in (\mathbf{P}_1, \mathbf{P}_2)$, and fixed $S \in \mathbf{T}$, we denote the expected value of L by

$$J_S(p_1, p_2) := E^{p_1, p_2}\{L(\gamma(\mathbf{x}); \beta(\mathbf{x}), \Theta_S)\}. \tag{4}$$

The problem, then, is to determine the **saddle point** of the game defined by the kernel J_S, *i.e.* to obtain a pair $(p_1^*, p_2^*) \in (\mathbf{P}_1, \mathbf{P}_2)$ such that, for all $(p_1, p_2) \in (\mathbf{P}_1, \mathbf{P}_2)$,

$$J_S(p_1^*, p_2) \leq J_S(p_1^*, p_2^*) \leq J_S(p_1, p_2^*). \tag{5}$$

Such a ***saddle point*** necessarily exists for every $S \in \mathbf{T}$, since the strategy spaces of the original game (*i.e.* Γ for *Player 1* and $(\mathbf{B}, \boldsymbol{\Theta}_S)$ for *Player 2*) are finite.[1]

One would also be interested in obtaining a characterization of the saddle-point solution in the smaller class of ***Markov policies*** with finite memory, if such a characterization exists. Roughly speaking, such policies have the property that their domain of definition (in particular the memory required for their realization) does not grow with t. Let m be a fixed nonnegative integer, and define the information set η_t to be[2]

$$\eta_t = \begin{cases} \{x_s\}_{s=t-m}^{t} & t > m \\ x^t & t \leq m \end{cases}.$$

---

[1] *Actually* $\mathbf{B}$ *is only countable, because* $\mathbf{Z}$ *is countable; but because of the structure of the objective function, and the fact that* $|x_t - x_1| < t$, *for all* t, V *(and hence* $\mathbf{B}$*) can be replaced by a finite set.*

[2] *Two such choices would be* $\eta_t = \{x_t\}$ *and* $\eta_t = \{x_t, x_{t-1}\}$, $t \in \mathbf{T}$, *which correspond to the "memoryless" and "one-step memory" cases, respectively.*

Then, a finite (m-step) memory Markov policy for *Player 1* (with respect to the information η_t) is a mapping $\hat{\gamma}_t : \eta_t \mapsto p_{ut} \in P(U)$, where $P(U)$ is the class of all probability measures on U.

A motivation for the above formulation is provided by the scenario of the so-called *Rabbit and Hunter game* (*Bernhard and Colomb, 1988*)[3]. For simplicity, let S=1; then the interpretation for the game goes as follows: A rabbit moves along a straight line which is made up of discrete slots (positions), and from one time instant to the next it can hop either to the right or to the left or stay where it is. A hunter, operating in the dark, attempts to shoot the rabbit by the end of T time units. He does not know the initial position of the rabbit, but has a flash light which he can use only once to acquire perfect information on the current position of the rabbit. At this point he shoots a single bullet which takes two time units to reach the target. For a general S, he can repeat this S times, but he is not allowed to carry information from one time instant to another. The objective function (*i.e.* (2*b*)) is the probability with which the rabbit is shot at the end of the T time units, which is to be maximized by the hunter (*Player 2*) and minimized by the rabbit (*Player 1*). Yet another motivation for the game could be provided by a "target location estimation" problem where the target again moves in a straight line, and the signal from the target (such as *sonar* reflecting from a submarine) takes 2 time units to reach the decision unit. A special version of the problem above, with the information delay (n) being 1 unit, was briefly discussed earlier by *Ferguson (1967)*, and a stochastic version, again with $n = 1$, was studied by *Bernhard, Colomb and Papavassilopoulos (1987)*.

The organization of the present paper is as follows. In *section 2* we consider the special case corresponding to $S = 1$, and obtain the complete set of (mixed) saddle-point solutions through a new approach which decomposes the original dynamic game into simpler, but *overlapping*, static subgames. In *section 3*, we generalize this solution to the case of arbitrary (but fixed) S, as formulated in *section 1*, and then we further extend the result to cover more general situations where the action set of *Player 1* has more than 3 elements and the delay factor in (2*a*) is more than 2 time units. An example included in the section serves to illustrate this extension. A second extension in *section 3* involves the situation where the action space of *Player 1* is instead the *plane*, for which we again construct a class of stationary Markov strategies which are in saddle-point equilibrium. *Section 4* introduces the game where *Player 2* is allowed to use memory strategies, it obtains the solution for an instance of the problem, with $T = 2$, $n = 2$, $m = 1$, and identifies a number of challenging open questions in attempts towards developing the counterparts of the results presented for the no-memory case. Finally, *section 5* includes a discussion on some possible formulations and solution methods for the "noisy" case, where *Player 2* does not have perfect knowledge of *Player 1*'s initial as well as current positions.

---

[3] *The scenario here is in fact somewhat different from the one discussed in that reference, since we do not allow for the hunter to observe the position of the rabbit at each point in time and thus construct a policy based on that memory information. Problems of this type, where v_t is allowed to depend also on past values of the state trajectory (i.e. on x^t, instead of x_t) will be discussed briefly in section 4.*

2. The Case S=1

We first consider the case $S = 1$, where we denote the singleton set Θ_1 by s, and rewrite the objective function as

$$L(\mathbf{u}; v_s, s) := \begin{cases} 1 & \text{if} \quad v_s = x_{s+2} \\ 0 & \text{otherwise} \end{cases} , \tag{6}$$

where $s \in \mathbf{T}$ is also a variable (whose probability distribution will be determined by *Player 2*). In principle, for every fixed T, this zero-sum game can be solved by first converting it into a matrix game and then computing a mixed saddle-point solution for the resulting matrix. In the construction of this matrix game, *Player 1*'s information can be taken to be *open-loop*, without any loss of generality, leaving him with $a_1 := 3^{T+2}$ alternatives, since $card(U) = 3$ and there are $T+2$ decision periods. For *Player 2*, on the other hand, for each s, we can restrict his choices to a set $\mathbf{B}_s$ of cardinality 5, with a typical element being in the form

$$\beta_s(x_s) = \{x_s + i; \;\; i \in \{-2, -1, 0, 1, 2\} \; \} ,$$

Since $card(\mathbf{T}) = T$, the total number of alternatives facing *Player 2* is $5T =: a_2$, and hence the matrix game is of dimensions $a_1 \times a_2$, which grows exponentially with T. This makes the computation of the mixed saddle-point solution quite prohibitive, even for relatively small values of T. We therefore abandon this brute-force computation. On the other hand, since the game is not a *feedback game* (because of the two-step delay information pattern), a temporal decomposition would not be possible, as in the derivation of *stagewise equilibria* (*Başar and Olsder, 1982*). Hence, ruling out these two standard approaches for derivation of mixed saddle-point equilibria, we now attempt to develop a new method for the solution of the dynamic game under study.

Toward the above goal, let us first consider a different (but related) game where *Player 1* is also endowed with the realized value of s; that is, the only difference between the original game and the new one is that a typical pure strategy for *Player 1* has been changed to $\gamma_t : (x^t, s) \mapsto u_t \in U$. Now, conditioned on s, the only relevant variables in this new game, that affect the objective function, are u_s, u_{s+1} and v_s, and the precise value of x_s does not play a role. Hence, apart from a change in notation, the new game is equivalent to the original one with $T = 1$, where necessarily $s = 1$. Therefore, we now direct our attention toward obtaining the saddle-point equilibria of the game for this special case (*i.e.* $T = 1$).

For $T = 1$, the relevant variables are x_3 and v_1, where both take values in the set $\{0, \pm1, \pm2\}$, and the objective is to determine two probability mass functions (*pmf*'s), one for x_3 and the other one for v_1, so that the former minimizes the probability of occurrence of the event $\{x_3 = v_1\}$ and the latter maximizes the same. It is fairly easy to see that this game (in terms of the variables x_3 and v_1) admits a **unique** saddle point, characterized by

$$Prob(x_3 = i) = Prob(v_1 = i) = \frac{1}{5}, \quad i = 0, \pm1, \pm2 , \tag{7}$$

where $Prob(\cdot) :=$ Probability of the event $\{\cdot\}$. The *value* of the game is $\frac{1}{5}$. This, however, does not completely characterize the solution to the original game with $T = 1$, since there the decision variables for *Player 1* were u_1 and u_2, and not x_3, with a many-to-one mapping from the former to the latter. Since $x_3 = u_1 + u_2$, the saddle-point solution to the original game (with $T = 1$) can easily be constructed from (7), but clearly it is not going to be unique. The complete solution is given in the following *Proposition*, for which we do not provide a proof here in view of the preceding discussion.

Proposition 2.1. *The zero-sum game of section 1, with $T = 1$, admits multiple saddle-point equilibria in mixed strategies, with a saddle-point value of 1/5. For Player 2, the saddle-point strategy is unique and is given by*

$$Prob(v_1 = i) = \frac{1}{5}, \quad i = 0, \pm 1, \pm 2. \tag{8a}$$

For Player 1, the complete set of saddle-point strategies is

$$\left.\begin{aligned} &P_{1,1} = P_{-1,-1} = \frac{1}{5}; \quad P_{1,0} = \frac{1}{5} - P_{0,1} = a \\ &P_{0,-1} = \frac{1}{5} - P_{-1,0} = b \\ &P_{1,-1} = c, \quad P_{-1,1} = d, \quad P_{0,0} = \frac{1}{5} - c - d \end{aligned}\right\} \tag{8b}$$

where $P_{i,j} := Prob(u_1 = i, u_2 = j)$, and

$$0 \le a,\ b,\ c,\ d \le \frac{1}{5}, \quad c + d \le \frac{1}{5}.$$

◇

Hence, the solution to the new game (where *Player 1* has also access to the value of s) is exactly the one given in *Proposition 2.1* above, with only v_1 in (8a) replaced by $v_s - x_s$, and in the definition of $P_{i,j}$ u_1 and u_2 replaced by u_s and u_{s+1}, respectively. Now, toward construction of a solution to the original game without the informational advantage to *Player 1* (and still with $S = 1$), let us suppose (for the time being) that the saddle-point solution dictates the probability distribution on $\mathbf{T}$ to be *one point*. If this assertion were true, then in solving for p_1^* from the right-hand-side (*RHS*) inequality of (5), *Player 1* would be faced with a game exactly of the type covered by *Proposition 2.1* (only possibly shifted in time). But this time, the game has to be solved for each s, since the actual value of s is not prior knowledge to *Player 1*. This means that *Player 1* has to determine a *pmf* for $\mathbf{u}$, such that for any frozen s, the pair (u_s, u_{s+1}) is a random sequence with the probability mass function (8b), where $P_{i,j}$ is replaced by

$$P_{i,j}^s := Prob(u_s = i, u_{s+1} = j).$$

Since the side condition involves a pair of random variables, the construction of a *pmf* for $\mathbf{u}$, that satisfies the given $T+1$ side conditions, may not always be possible because of the strong coupling (or *overlap*) across consecutive stages. One has to verify a **consistency condition** which basically says:

For two "neighboring" sequences (u_s, u_{s+1}) and (u_{s+1}, u_{s+2}), both having probability mass functions as given by (8b), the u_{s+1} appearing in the two sequences should be the same random variable.

Another equivalent statement would be the following:

There exists a probability mass function for the triple (u_s, u_{s+1}, u_{s+2}), whose projections onto (u_s, u_{s+1}) and (u_{s+1}, u_{s+2}) are consistent with the given probability mass functions for these two sequences - this being true for all $s \in \mathbf{T}$.

One easy way to verify the existence of such a distribution would be to check the *consistency* of the marginals for u_{s+1}, obtained from (u_s, u_{s+1}) and (u_{s+1}, u_{s+2}). Proceeding in this manner, we require (using (8b)):

$$\sum_{i=-1}^{1} P_{i,j}^{s} \equiv \sum_{i=-1}^{1} P_{j,i}^{s+1} \ , \quad j = 0, \pm 1 \ , \tag{9a}$$

which holds if, and only if,

$$a = b = 0.1 \ , \quad c = d = \alpha \ , \quad \alpha \in [0, 0.1] \ . \tag{9b}$$

Hence, there exist uncountably many such distributions. For each fixed $\alpha \in [0, 0.1]$, the corresponding probability mass function for **u** can be obtained recursively from

$$Prob(u_{t+1} = i \mid u_t = j) = P_{i|j}^{*}(\alpha)\ ; \quad Prob(u_t = i) = P_i^{*}(\alpha), \ \ t \leq T + 1 \ , \tag{10a}$$

where (suppressing the dependence on α):

$$\left.\begin{aligned}
&P_1^* = P_{-1}^* = 0.3 + \alpha\ ; \quad P_0^* = 0.4 - 2\alpha \\
&P_{1|1}^* = P_{-1|-1}^* = \frac{0.2}{0.3+\alpha}\ ; \quad P_{1|-1}^* = P_{-1|1}^* = \frac{\alpha}{0.3+\alpha} \\
&P_{0|0}^* = \frac{0.2-2\alpha}{0.4-2\alpha}\ ; \quad P_{-1|0}^* = P_{1|0}^* = \frac{0.1}{0.4-2\alpha} \\
&P_{0|1}^* = P_{0|-1}^* = \frac{0.1}{0.3+\alpha}\ .
\end{aligned}\right\} \tag{10b}$$

This makes the sequence $\{u_t\}$ a Markov chain with one-step memory, whose transition diagram is depicted in *Fig. 1*, on the next page, for two extreme values of α, 0 and 0.1, and one intermediate value, 1/30. The significance of the chosen intermediate value, $\alpha = 1/30$, is that the corresponding Markov chain visits the three states, $1, 0, -1$, with equal probability, 1/3.

This now brings us to the following main result of this section.

Theorem 2.1. *The zero-sum dynamic game of this section (with $S = 1$) admits a saddle-point solution in mixed strategies, given as follows:*

i. *The mixed saddle-point strategy for Player 1, p_1^*, is a probability mass function for* **u**, *which defines a stationary Markov chain with transition probabilities*

$$Prob(u_{t+1} = i \mid u_t = j) = P^*_{i|j}(\alpha)\,, \quad t \geq 1\,, \tag{11a}$$

$$Prob(u_1 = i) = P^*_i(\alpha) \tag{11b}$$

as given by (9), *for each $\alpha \in [0,\ 0.1]$.*

ii. *The mixed saddle-point strategy for Player 2, p_2^*, has two components: The variable s is taken as a deterministic quantity, to be any element of* **T**. *Given the specific choice for s, and the observed value of x_s, the action variable v_s is taken as a random variable with support $x_s + \{0, \pm 1, \pm 2\}$, and with each of the five points receiving equal probability, 1/5.*

iii. *The saddle-point value of the game (in mixed strategies) is* 1/5.

Proof. Given p_2^*, as in *(ii)* above, *Player 1* faces a game with only two stages, as discussed prior to *Proposition 2.1*. Then, his best response is given by (8b) in *Proposition 2.1*, for which a valid representation in the original strategy space is (11) above, since the Markov chain has the two-step invariant distribution (8b) with $a = b = 0.1$, $c = d = \alpha$, as verified prior to the statement of the Theorem.

To verify the *LHS* inequality of (5), we take p_1^* as defined by (11) and observe that the maximization problem faced by *Player 2* is independent of s; hence, any value for s (including any distribution on **T**) would be a valid (optimizing) choice. But once s is fixed (frozen), the problem is the one covered by *Proposition 2.1*, from which the optimality of p_2^* follows. Consequently, the saddle-point values of the two games covered by *Proposition 2.1* and *Theorem 2.1* are the same. ◇

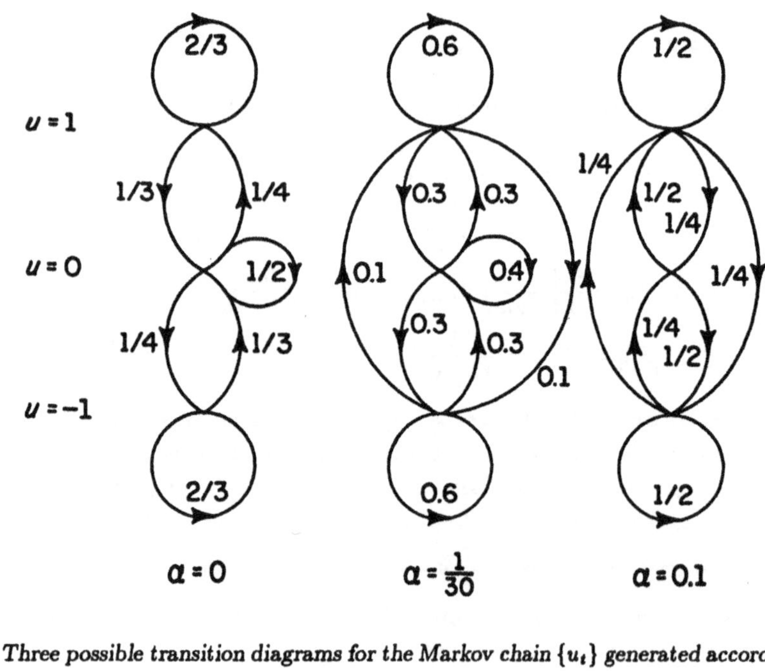

Fig. 1: Three possible transition diagrams for the Markov chain $\{u_t\}$ generated according to (10).

Remark 2.1. *Theorem 2.1* above does not cover all possible saddle points of the game (with $S = 1$). As hinted in the proof of the *Theorem*, one could in fact choose s to be any random variable on the finite set $\mathbf{T}$. The *LHS* inequality of (5) holds because under p_1^* the maximization problem faced by *Player 2* is independent of s (and also of v_s on the set $x_s + \{0, \pm 1, \pm 2\}$), and the *RHS* inequality holds because under p_2^* the minimization problem faced by *Player 1* is independent of $\{x_t\}$ (and hence of $\{u_t\}$). ◇

3. The General Case and Two Extensions

The symmetric nature of the solution presented in *Theorem 2.1* for the case $S = 1$ readily lends itself into a candidate solution for the general case when S is any element of $\mathbf{T}$. In particular, under the p_1^* given in *Theorem 2.1*, *Player 2*'s maximization problem is independent of v_t on the set $\{x_t + i\,;\ i = 0, \pm 1, \pm 2\}$, for all $t \in \mathbf{T}$, as we argued earlier. Hence, any choice for Θ_S, along with any probability distribution on the set $\{x_t + i\,;\ i = 0, \pm 1, \pm 2\}$, for $t \in \Theta_S$, will return the value of $\frac{1}{5}S$ to *Player 2*. Accordingly, we take p_2^* as a probability mass function that assigns any probability distribution to the elements of $\boldsymbol{\Theta}_S$, and for the support set of any such realization, say Θ_S, it assigns equal probability to the five points $\{x_t + i\,;\ i = 0, \pm 1, \pm 2\}$ for every $t \in \Theta_S$. Under such a choice of p_2^*, the minimization problem faced by *Player 1* is again independent of $\{x_t\}$, thus making the distribution given by (10) still an acceptable choice for any $\alpha \in [0, 0.1]$. This then leads to the following *Theorem* which provides the complete solution to the game initially formulated in *section 1*.

Theorem 3.1. *The zero-sum dynamic game originally formulated in section 1 admits multiple saddle-point equilibria, given as follows:*

i. For Player 1, any Markov chain described by (11) *in Theorem 2.1 constitutes a saddle-point strategy.*

ii. For Player 2, the mixed saddle-point strategy p_2^* has two components:

(a) Any probability distribution on the class of sets $\boldsymbol{\Theta}_S$;

(b) For a given realization on $\boldsymbol{\Theta}_S$, say Θ_S, a sequence of S random variables $\{v_t\,,\ t \in \Theta_S\}$ where v_t takes the values $x_t + \{0, \pm 1, \pm 2\}$, each with equal probability $1/5$.

iii. The saddle-point value (in mixed strategies) is $S/5$. ◇

Remark 3.1. Note the crucial assumption that *Player 2* is not allowed to use memory in the construction of his strategies. If this were not the case, the statement of *Theorem 3.1* would not necessarily have been true, because then *Player 2* will not be facing identical maximization problems for different $t \in \Theta_S$. Further discussion on this point, including the derivation of some saddle-point policies with memory, can be found in the next section. ◇

AN EXTENSION

One could envisage an extension of the main result given above to the more general scenario where

(i) the action set of *Player 1* is replaced by $U = \{0, \pm 1, .., \pm m\}$ where m is a positive integer, and

(ii) the function in (2a) is replaced by

$$L_t := \begin{cases} 1 & \text{if } v_t = x_{t+n} \\ 0 & \text{otherwise} \end{cases} \tag{12}$$

where $n \geq 2$ is also an integer.
Following the basic idea that led to *Theorems 2.1* and *3.1*, we first consider the case $S = 1$, and freeze the variable s, thus arriving at a game of the type covered by *Proposition 2.1*, with the only difference being that the action space of *Player 1* is now a much larger set. Note that *Player 2* acts in this special game only once, with the relevant part of his action space being $V = \{0, \pm 1, .., \pm mn\}$. With x_{n+1} taken as the (aggregate) decision variable of *Player 1*, this game admits a unique saddle-point solution, where

$$Prob(x_{n+1} = i) = \frac{1}{2mn+1}, \quad i \in V \tag{13a}$$

$$Prob(v_1 = i) = \frac{1}{2mn+1}, \quad i \in V \ , \tag{13b}$$

and the saddle-point value is $1/(2mn+1)$. In terms of the variables of the original game (with $T = 1$), *i.e.* $u_1, .., u_n$, the saddle-point policy for *Player 1* is any probability distribution for u^n on $\mathbf{X}_{j=1}^n U$, satisfying the side condition that the variable $x_{n+1} := \sum_{j=1}^n u_j$ has the probability mass function (13a). Let $\mathbf{P}^{(n)}$ be the class of all such *pmf's* (for *Player 1*) on $\mathbf{X}_{j=1}^n U$, which satisfy the given side condition. Denote a typical element of $\mathbf{P}^{(n)}$ by $\{P^{(n)}_{i_1,..,i_n}\}$, each term here admitting the interpretation

$$P^{(n)}_{i_1,..,i_n} := Prob(u_1 = i_1, u_2 = i_2, .., u_n = i_n). \tag{14a}$$

For each such probability mass function, and given two integers j, k $(1 \leq j \leq k \leq n)$, we introduce the associated $(k-j+1)$-dimensional marginal probability mass function

$$P^{(k-j+1)}_{i_j,..,i_k} := \sum_{i_1,..i_{j-1}} \sum_{i_{k+1},..,i_n} P^{(n)}_{i_1,..,i_n} = Prob(u_j = i_j, .., u_k = i_k). \tag{14b}$$

We now require the following condition to hold for $\{P^{(n)}_{i_1,..,i_n}\}$, as the counterpart of (9a) in this case:

Consistency Condition. *The probability mass functions* $\{P^{(n)}_{i_1,..,i_n}\}$ *have the property that, for every integer* j, $2 \leq j \leq n$,

$$\{P^{(n-j+1)}_{i_j,..,i_n}\} \equiv \{P^{(n-j+1)}_{i_1,..,i_{n-j+1}}\},$$

which is a way of saying that the marginal distributions of the two sequences $(u_j, .., u_n)$ *and* $(u_1, .., u_{n-j+1})$ *are identical for every integer* j, $2 \leq j \leq n$. ◇

We now note that there is always a *pmf* $\{P^{(n)}_{i_1,..,i_n}\}$ that satisfies the *Consistency Condition* along with the side condition (13a). One such example is the *completely symmetric pmf, i.e.* the *pmf* with the property that

$$P^{(n)}_{i_1,..,i_n} \equiv P^{(n)}_{i'_1,..,i'_n}$$

where $\{i'_1,..,i'_n\}$ is any arrangement of $\{i_1,..,i_n\}$. Such a *pmf* can always be constructed under the side condition (13a) since the side condition itself is symmetric. Note that the *pmf* given by (8b) is completely symmetric for precisely the set of values of a, b, c, d (given by (9b)) that make (9a) be satisfied.

In principle, it is possible to construct the entire family of *pmf*'s with the above property, but for general n and m such a construction would be notationally cumbersome, and hence it will not be included here. What we can say, however, is that the Markov chain associated with such a *pmf* will not generally admit the simple structure (11); instead the transitions will involve an $(n-1)$-step memory, *i.e.* for a complete description what we will need is

$$P_{i_n|i_1,..,i_{n-1}} := Prob(u_t = i_n \mid u_{t-n+1} = i_1, .., u_{t-1} = i_{n-1}) ,$$

to replace (11a). This, along with $\{P^{(n-1)}_{i_1,..,i_{n-1}}\}$, generates a valid *pmf* for the entire sequence $\{u_1, .., u_{T+n-1}\}$.

Clearly, the *Consistency Condition* ensures that the solution of the game with $T = 1$ can be extended to cover the game with $S = 1$ (but $T > 1$), as in the case of *Theorem 2.1*. Likewise, a further extension from the game with $S = 1$ to the more general game with $S > 1$ is possible along the directions that led us to proceed from *Theorem 2.1* to *Theorem 3.1*. The final result, then, is the one given in *Theorem 3.2* below, which however is not as precise as *Theorem 3.1* is because we have not constructed explicitly the underlying probability distributions for *Player 1*.

Theorem 3.2. *The more general zero-sum dynamic game, where* $U = \{0, \pm 1, .., \pm m\}$ *and* L_t *is defined by* (12), *admits a saddle-point solution in mixed strategies, given as follows:*

i. *For Player 1, any time-invariant distribution for* u^{T+n-1}, *that satisfies the Consistency Condition and the side condition* (13a), *is a saddle-point strategy. Such a probability distribution exists.*
ii. *For Player 2, the saddle-point strategy is as in Theorem 3.1, with the only difference being that now the variable* v_t, $t \in \Theta_S$, *takes the values* $x_t + \{0, \pm 1, .., \pm mn\}$, *each with equal probability* $1/(2mn+1)$.
iii. *The saddle-point value of the game (in mixed strategies) is* $S/(2mn+1)$. ◇

An Example. To illustrate the result of *Theorem 3.2*, let us consider a specific example where $m = 1, n = 3$, and S, T are arbitrary.

The side condition (13a) in this case is

$$Prob(u_1 + u_2 + u_3 = i) = \frac{1}{7}, \quad i = 0, \pm 1, \pm 2, \pm 3.$$

A *pmf* for (u_1, u_2, u_3), which meets the above condition and also the *Consistency Condition*, is $\{P_{i,j,k}\}$, where

$$P_{i,i,i} = P_{0,0,0} = \frac{1}{7}, \quad i = \pm 1$$

$$P_{i,0,i} = P_{0,i,i} = P_{i,i,0} = P_{i,0,0} = P_{0,i,0} = P_{0,0,i} = \frac{1}{21}, \ i = \pm 1 ,$$

and other entries are *zero*.
This is equivalently expressed in terms of the conditional probabilities $P_{i|j,k} := Prob(u_3 = i \mid u_1 = j, u_2 = k)$:

$$P_{i|i,i} = \frac{3}{4}, \quad i = \pm 1$$

$$P_{i|i,0} = P_{i|0,i} = P_{0|0,i} = P_{0|i,0} = \frac{1}{2}, \ i = \pm 1$$

$$P_{i|0,0} = \frac{1}{5}, \quad P_{0|i,i} = \frac{1}{4}, \ i = \pm 1$$

$$P_{i|i,-i} = P_{i|-i,i} = P_{i|-i,-i} = P_{0|i,-i} = 0 , \ i = \pm 1$$

along with the starting probabilities for (u_0, u_1):

$$P_{i,i} = \frac{4}{21}, \ P_{i,0} = P_{0,i} = \frac{2}{21}, \quad i = \pm 1$$

$$P_{0,0} = \frac{5}{21}, \ P_{i,-i} = 0 , \quad i = \pm 1 .$$

The above are, in fact, the transition probabilities that generate the *pmf* for the entire sequence u^{T+2}, where for any $t \geq 1$

$$Prob(u_{t+2} = i \mid u_t = j, u_{t+1} = k) = P_{i|j,k} .$$

It is easy to see that $P_{i|j,k} \neq P_{i|k}$, which means that the saddle-point strategy for *Player 1* cannot be generated (at least in the class considered) using only one-step memory transition probabilities, thus corroborating our earlier remark. ◇

A SECOND EXTENSION

We now consider a second extension on the basic model of *section 2*, where the real line ($\mathbf{R}$) is replaced by the plane ($\mathbf{R}^2$). This is the case where *Player 1* is allowed to move on the plane, to one of five possible positions at each point in time; more precisely, visualizing a rectangular mesh on the plane, at any point t *Player 1* is in one of the cells and is allowed either to stay where he is or to make a move in horizontal (up or down) or vertical (right or left) directions. We now have two-dimensional representations for the state and the action variables: $x := (x^1, x^2)$, $u := (u^1, u^2)$, $v := (v^1, v^2)$, where the components of x and v (denoted by superscript indices) still belong to $\mathbf{Z}$, and u takes values in $U := \{(-1,0),(0,-1),(0,0),(0,1),(1,0)\}$. Here we have adopted the convention that the first (respectively, second) component designates the position or move in the horizontal (respectively, vertical) direction. The basic state equation (1) remains intact, with the understanding that we now have componentwise addition; furthermore, the loss function L is still as defined in (2), with the equality in (2a) again to be interpreted in the vector sense.

We now claim that the approach adopted in *section 2* equally applies to this problem, and a saddle-point solution can be constructed explicitly. To substantiate this claim, we simply produce, for the discrete game on the plane, the counterparts of *Proposition 2.1*, the *consistency condition* (9), and *Theorem 2.1*, to be called *Proposition 3.1*, *Consistency Condition ("plane")*, and *Theorem 3.3*, respectively.

Proposition 3.1. *The zero-sum discrete game on the plane, with $T = 1$, admits multiple saddle-point equilibria in mixed strategies, with a saddle-point value of 1/13. For Player 2, the saddle-point strategy is unique and is given by*

$$Prob(v_1 = (i,j)) = \frac{1}{13}, \quad |i| + |j| \leq 2. \tag{15a}$$

For Player 1, the complete set of saddle-point strategies is

$$\left.\begin{array}{l}
P_{(0,-1),(0,-1)} = P_{(0,1),(0,1)} = P_{(1,0),(1,0)} = P_{(-1,0),(-1,0)} = \frac{1}{13} \\
P_{(0,1),(1,0)} = \frac{1}{13} - P_{(1,0),(0,1)} = a_1 \;; \quad P_{(0,1),(-1,0)} = \frac{1}{13} - P_{(-1,0),(0,1)} = a_2 \\
P_{(0,-1),(1,0)} = \frac{1}{13} - P_{(1,0),(0,-1)} = a_3 \;; \quad P_{(0,-1),(-1,0)} = \frac{1}{13} - P_{(-1,0),(0,-1)} = a_4 \\
P_{(0,0),(0,1)} = \frac{1}{13} - P_{(0,1),(0,0)} = b_1 \;; \quad P_{(0,0),(1,0)} = \frac{1}{13} - P_{(1,0),(0,0)} = b_2 \\
P_{(0,0),(0,-1)} = \frac{1}{13} - P_{(0,-1),(0,0)} = b_3 \;; \quad P_{(0,0),(-1,0)} = \frac{1}{13} - P_{(-1,0),(0,0)} = b_4 \\
P_{(0,1),(0,-1)} = c_1 \,, \quad P_{(0,-1),(0,1)} = c_2 \,, \quad P_{(1,0),(-1,0)} = c_3 \,, \quad P_{(-1,0),(1,0)} = c_4 \\
P_{(0,0),(0,0)} = \frac{1}{13} - c_1 - c_2 - c_3 - c_4
\end{array}\right\} \tag{15b}$$

where $P_{(i_1,i_2),(j_1,j_2)} := Prob(u_1 = (i_1,i_2), u_2 = (j_1,j_2))$, and

$$0 \leq a_i,\; b_i,\; c_i \leq \frac{1}{13}, \quad c_1 + c_2 + c_3 + c_4 \leq \frac{1}{13}.$$

Consistency Condition ("plane"). *The probability mass functions*

$$\{P^s_{(i_1,i_2),(j_1,j_2)} := Prob(u_s = (i_1,i_2), u_{s+1} = (j_1,j_2))\}$$

have the property that, for every $(j_1,j_2) \in U$,

$$\sum_{(i_1,i_2)\in U} P^s_{(i_1,i_2),(j_1,j_2)} = \sum_{(i_1,i_2)\in U} P^{s+1}_{(j_1,j_2),(i_1,i_2)}$$

Again there exist uncountably many *pmf*'s that satisfy the *consistency condition*. In particular, choosing $a_i = b_i = 1/26, i = 1,...,4$; $c_1 = c_2 = \xi$, $c_3 = c_4 = \beta$, one can generate a class of *pmf*'s which satisfy the condition for all nonnegative ξ, β, $\xi + \beta \leq 1/26$. In mimmicking the transition diagrams of *Fig. 1*, we have $\xi = \beta = 0$ corresponding to $\alpha = 0$ there, which corresponds to the scenario when two consecutive moves along the horizontal or vertical directions are not allowed. The other extreme case corresponds to the set of values $\xi + \beta = 1/26$, which indicates that *Player 1* cannot "take a rest" for two consecutive periods. Finally, for the values $\xi + \beta = 1/65$, all five states of the underlying Markov chain receive equal residence time. For general ξ and β, the *pmf* for **u** can be generated recursively from

$$Prob(u_{t+1} = (i_1, i_2) \mid u_t = (j_1, j_2)) = P^*_{i_1,i_2|j_1,j_2}(\xi, \beta);\ Prob(u_t = (i_1, i_2)) = P^*_{i_1,i_2}(\xi, \beta)\,, \quad (16a)$$

where (suppressing the dependence on ξ and β):

$$\left.\begin{aligned}
&P^*_{1,0} = P^*_{-1,0} = \frac{5}{26} + \beta\,; \quad P^*_{0,1} = P^*_{0,-1} = \frac{5}{26} + \xi\,; \quad P^*_{0,0} = \frac{3}{13} - 2(\xi + \beta) \\
&P^*_{1,0|1,0} = P^*_{-1,0|-1,0} = \frac{2}{5 + 2\beta}\,; \quad P^*_{0,-1|0,-1} = P^*_{0,1|01} = \frac{2}{5 + 2\xi} \\
&P^*_{1,0|0,0} = P^*_{-1,0|0,0} = P^*_{0,-1|0,0} = P^*_{0,1|0,0} = \frac{1}{6 - 4(\xi + \beta)} \\
&P^*_{-1,0|1,0} = P^*_{1,0|-1,0} = \frac{26\beta}{5 + 26\beta} \\
&P^*_{0,0|0,-1} = P^*_{0,0|0,1} = \frac{1}{5 + 26\xi}\,; \quad P^*_{0,0|-1,0} = P^*_{0,0|1,0} = \frac{1}{5 + 26\beta} \\
&P^*_{0,0|0,0} = \frac{1 - 26(\xi + \beta)}{3 - 2(\xi + \beta)}\,; \quad P^*_{0,-1|0,1} = P^*_{0,1|0,-1} = \frac{26\xi}{5 + 26\beta} \\
&P^*_{1,0|0,1} = P^*_{-1,0|0,1} = P^*_{1,0|0,-1} = P^*_{-1,0|0,1} = \frac{1}{5 + 26\xi} \\
&P^*_{0,-1|-1,0} = P^*_{0,1|-1,0} = P^*_{0,1|1,0} = P^*_{0,-1|1,0} = \frac{1}{5 + 26\beta}
\end{aligned}\right\} \quad (16b)$$

Theorem 3.3. *The zero-sum discrete dynamic game defined on the plane (with $S = 1$), as above, admits a saddle-point solution in mixed strategies, given as follows:*

*i. The mixed saddle-point strategy for Player 1, $\mathbf{p}_1^*$, is a probability mass function for **u**, which defines a stationary Markov chain with transition probabilities*

$$Prob(u_{t+1} = (i_1, i_2) \mid u_t = (j_1, j_2)) = P^*_{i_1,i_2|j_1,j_2}(\xi, \beta)\,, \quad t \geq 1\,, \quad (17a)$$

$$Prob(u_1 = (i_1, i_2)) = P^*_{i_1,i_2}(\xi, \beta)\,, \quad (17b)$$

as given above, for each nonnegative ξ and β such that $\xi + \beta \leq 1/26$.

ii. The mixed saddle-point strategy for Player 2, $\mathbf{p}_2^*$, has two components: The variable s is taken as a deterministic quantity, to be any element of $\mathbf{T}$. Given the specific choice for s, and the observed value of x_s, the action variable v_s is taken as a random variable with support $x_s + U + U$, and with each of the thirteen points receiving equal probability, 1/13.

iii. The saddle-point value of the game (in mixed strategies) is 1/13. ◇

Remark 3.2. Since one can set up a one-to-one correspondence between cells on the plane and integers on the real line, one may wonder whether *Theorem 3.3* would be subsumed by *Theorem 3.2*, for some particular value of m. This turns out not to be the case because it is not possible to carry moves on the plane to moves on the real line in a linear fashion, so as to preserve the original neighboring relationships. This latter feature is essential in the verification of the consistency condition and existence of stationary Markov policies. It is also worth pointing to the fact that under any transformation that maps the plane onto the real line, the cardinality of the action set has to grow with time, and hence no formulation on the real line, with a fixed action set, would be able to capture the dynamics on the plane. A final note is that the result of *Theorem 3.3* extends naturally to the more general case when delay, n, is more than 2 time units, and moves along the horizontal and vertical directions are allowed to be more than one cell at a time. An explicit construction of the saddle-point solution, however, for general n and m, is notationally quite prohibitive. ◇

4. Games with Memory Strategies for Player 2

In this section we relax one of the major assumptions of the previous sections, namely the memoryless nature of the strategies for *Player 2*. In a way, this is the "day-light" version of the earlier games, where the maximizer is allowed to make perfect measurements on the current positions of the minimizer, including the initial state, and recall this position information throughout the duration of the game. Hence, a typical position-forecast strategy for *Player 2* will now be a function $\beta_t : x^t \mapsto v_t \in V$, and otherwise the game is identical with the one formulated in *section 1*. Here, we discuss only the case $S = 1$ and particularly explore the possibility of using the technique of *section 2* for the problem at hand.

Clearly, *Player 2* cannot do any worse in this new game, than the saddle-point value of *Theorem 2.1*, since he has additional information on the past positions of *Player 1*. The question then is whether there would exist strategies for *Player 1* under which *Player 2* would still be forced to pick the strategy given in *Theorem 2.1(ii)*. Since the underlying game is finite, it still admits a saddle point (in mixed policies), and furthermore if any saddle-point policy for *Player 2* dictates a probability-one choice for a particular value of s, then *Proposition 2.1* would still be relevant and the saddle-point value of the game will be 1/5. However, the *consistency condition* of *section 2* will have to be modified now, in view of the growing information set of *Player 2*. We now require the consistency of conditional probability mass functions (conditioned on the current and past state information) rather than that of unconditional *pmf*s as in *section 2*. Toward a precise mathematical statement of the underlying condition here, introduce the conditional *pmf*

$$P^s_{i,j}(x^t) := Prob(u_s = i, u_{s+1} = j | x^t) \,.$$

Then, the *consistency condition* that relates the *pmf*s at the time instants t and $t+1$ is:

$$\sum_j P_{i,j}^{t+1}(x^t, u_t = k) = \frac{P_{k,i}^t(x^t)}{\sum_{j'} P_{k,j'}^t(x^t)}, \quad k \in U \tag{18a}$$

which is to be used along with the side condition (from (8b)):

$$\left.\begin{array}{l} P_{1,1}^t(x^t) = P_{-1,-1}^t(x^t) = 0.2\,; \quad P_{1,0}^t(x^t) = 0.2 - P_{0,1}^t(x^t) = a_t(x^t) \\ P_{0,-1}^t(x^t) = 0.2 - P_{-1,0}^t(x^t) = b_t(x^t) \\ P_{1,-1}^t(x^t) = c_t(x^t)\,, \quad P_{-1,1}^t(x^t) = d_t(x^t)\,, \quad P_{0,0}^t(x^t) = 0.2 - c_t(x^t) - d_t(x^t) \end{array}\right\} \tag{18b}$$

where, for all reachable x^t and for all $t \leq T$,

$$0 \leq a_t(x^t),\ b_t(x^t),\ c_t(x^t),\ d_t(x^t) \leq 0.2\,, \quad c_t(x^t) + d_t(x^t) \leq 0.2\,. \tag{18c}$$

Note that in the above we allow for the possibility that the parameters a, b, c and d which were used in the construction of the saddle-point solution of *Theorem 2.1* may depend on both t and the history on the state. The earlier question can now be rephrased as the existence of functions $a_t(\cdot)$, $b_t(\cdot)$, $c_t(\cdot)$ and $d_t(\cdot)$ so that (18) is satisfied for all $t \leq T$. We show below that one can find a solution to (18) when $T = 1$ or $T = 2$, but no solution exists for $T > 3$. Henceforth, we take $x^1 \equiv x_1 = 0$, without any loss of generality.

Some algebra shows that the sequence $\{a_t(\cdot), b_t(\cdot), c_t(\cdot), d_t(\cdot);\ t \leq T\}$ satisfies (18) if, and only if, there exists a sequence of functions $\{m_t(\cdot), n_t(\cdot);\ t \leq T\}$:

$$m_t(x^t) = a_t(x^t) + c_t(x^t)\,, \quad n_t(x^t) = b_t(x^t) - d_t(x^t) \tag{19a}$$

$$0 \leq a_t, b_t, c_t, d_t \leq 0.2, \quad c_t + d_t \leq 0.2\,; \tag{19b}$$

which are generated through the recursions

$$\left.\begin{array}{r} m_{t+1}(x^t, 1) = \dfrac{0.2}{0.2 + m_t(x^t)} - 0.2 \\[2ex] m_{t+1}(x^t, -1) = \dfrac{d_t(x^t)}{0.4 - n_t(x^t)} - 0.2 \\[2ex] m_{t+1}(x^t, 0) = \dfrac{0.2 - a_t(x^t)}{0.4 - m_t(x^t) + n_t(x^t)} - 0.2 \end{array}\right\} \tag{20a}$$

$$\left.\begin{array}{r} n_{t+1}(x^t, -1) = 0.4 - \dfrac{0.2}{0.4 - n_t(x^t)} \\[2ex] n_{t+1}(x^t, 1) = 0.4 - \dfrac{c_t(x^t)}{0.2 + m_t(x^t)} \\[2ex] n_{t+1}(x^t, 0) = 0.4 - \dfrac{b_t(x^t)}{0.4 - m_t(x^t) + n_t(x^t)} \end{array}\right\} \tag{20b}$$

subject to the initial conditions

$$m_1(0) = a_1 + c_1\ , \quad n_1(0) = b_1 - d_1, \tag{20c}$$

where the quadruple (a_1, b_1, c_1, d_1) is determined so that

$$0 \le a_1, b_1, c_1, d_1 \le 0.2, \quad c_1 + d_1 \le 0.2. \tag{20d}$$

We now show that the above is a feasible set for $T \le 2$, but it does not admit a solution for $T > 3$. Towards this end we first choose

$$a_1 = b_1 = c_1 = d_1 = 0.1 \Longrightarrow m_1 = 0.2,\ n_1 = 0.$$

Then, using (20a) and (20b), we obtain

$$m_2(1) = 0.3, m_2(-1) = 0.05, m_2(0) = 0.3\ ; \quad n_2(-1) = -0.1, n_2(1) = 0.15, n_2(0) = -0.1,$$

which lead (using (19)) to more than one set of feasible values (in fact, an infinite number of them) for the quadruple $\{a_2(\cdot), b_2(\cdot), c_2(\cdot), d_2(\cdot)\}$. Two such solution sets are

$$\left.\begin{array}{l} a_2(1) = 0.20, b_2(1) = 0.15, c_2(1) = 0.10, d_2(1) = 0 \\ a_2(-1) = 0.05, b_2(-1) = 0, c_2(-1) = 0, d_2(-1) = 0.1 \\ a_2(0) = 0.20, b_2(0) = 0, c_2(0) = 0.10, d_2(0) = 0.1 \end{array}\right\} \tag{21a}$$

and

$$\left.\begin{array}{l} a_2(1) = 0.15, b_2(1) = 0.20, c_2(1) = 0.15, d_2(1) = 0.05 \\ a_2(-1) = 0.05, b_2(-1) = 0, c_2(-1) = 0, d_2(-1) = 0.1 \\ a_2(0) = 0.15, b_2(0) = 0, c_2(0) = 0.15, d_2(0) = 0.1 \end{array}\right\} \tag{21b}$$

Note that both of these lead to a conditional probability mass function $P^2_{i,j}(x^2)$, through (18b), that explicitly depends on $x^2 \in \{-1, 0, 1\}$. In fact, it is possible to show that every solution to the problem shows explicit dependence on x^2. The important point here is that, by using (only) nonstationary policies, the minimizer can still ascertain that the saddle-point value of $1/5$ is attained, in spite of the extra position information that the maximizer is endowed with, this being true (of course) for $T \le 2$. We now formalize this result in the following Proposition.

Proposition 4.1. *The zero-sum dynamic game of this section, with $T = 2$ (and $S = 1$ and the extra position information for Player 2) admits a saddle-point solution in mixed strategies, given as follows:*

i. The mixed saddle-point strategy for Player 1, p_1^*, is a pmf for $\mathbf{u}:=(u_1, u_2, u_3)$, which is generated by

$$Prob(u_2 = i, u_3 = j \mid u_1 \equiv x^2 = k) = P^2_{i,j}(k)\ , \quad i, j, k = 1, 0, -1\ , \tag{22a}$$

$$Prob(u_1 = 1) = Prob(u_1 = -1) = 2Prob(u_1 = 0) = 0.4 \tag{22b}$$

where $\{P^2_{i,j}(k)\}$ is determined from (18b) using the solution(s) to (19). Two such sets of solutions are given by (21a) and (21b).

ii. *The mixed saddle-point strategy for Player 2, p_2^*, has two components: The variable s is taken as a deterministic quantity, to be either 1 or 2. Given the specific choice for s, and the observed value of x_s, the action variable v_s is taken as a random variable with support $x_s + \{0, \pm 1, \pm 2\}$, and with each of the five points receiving equal probability, 1/5.*

iii. *The saddle-point value of the game (in mixed strategies) is 1/5.*

Proof. Given p_2^*, as in *(ii)* above, *Player 1* faces a game with only two stages, as discussed prior to *Proposition 2.1*. Then, his best response is given by (8b) in *Proposition 2.1*, for which a valid representation in the original strategy space is (22) above, since we also have $P^1_{k,i} = 0.1$ for all $k, i = 1, 0, -1$, except for $k = i = 0$ in which case it is 0.

To verify the *LHS* inequality of (5), we take p_1^* as defined by (22) and note that $Prob(u_1 + u_2 = i | x_1 = 0) = Prob(u_2 + u_3 = i | x_2 = j, x_1 = 0) = 0.2$ for all $i = 0, \pm 1, \pm 2,\ j = 0, \pm 1$. This implies that the maximization problem faced by *Player 2* is independent of s; hence, any value for s (including any distribution on $\mathbf{T}$) would be a valid (optimizing) choice. But once s is fixed (frozen), the problem is the one covered by *Proposition 2.1*, from which the optimality of p_2^* follows. Consequently, the saddle-point values of the two games covered by *Proposition 2.1* and *Proposition 4.1* are the same. ◇

For $T > 3$, however, a result of the type given in *Proposition 4.1* above does not hold, mainly because the set of equations (20) does not admit a solution. To prove the latter claim, assume (to the contrary) that there exists a solution to (20) for $T > 3$. Then, we first obtain from (20a), for $t \le T - 2$,

$$m_{t+2}(x^t, -1, 1) = \frac{0.2}{0.2 + m_{t+1}(x^t, -1)} - 0.2$$
$$= \frac{0.2}{d_t(x^t)}[0.4 - b_t(x^t)] \le 0.4$$

where the inequality holds since $m_{t+2} = a_{t+2} + c_{t+2}$, $0 \le c_{t+2} \le 0.2$ and $0 \le a_{t+2} \le 0.2$. This implies that

$$d_t(x^t) \ge 0.2 - \frac{1}{2} b_t(x^t) \ge 0.1, \quad t \le T - 2,$$

with the last step being due to the given upper bound on $b_t(x^t)$. In view of this,

$$n_t(x^t) := b_t(x^t) - d_t(x^t) \le 0.1, \quad t \le T - 2, \tag{$\bullet$}$$

and furthermore,

$$c_t(x^t) \le 0.2 - d_t(x^t) \le 0.1, \quad t \le T - 2. \tag{$\star$}$$

Now, working with (20b), we obtain, for all $t \le T - 2$,

$$n_{t+1}(x^t,1) = 0.4 - \frac{c_t(x^t)}{0.2+m_t(x^t)} \equiv \frac{0.2+a_t(x^t)}{0.2+a_t(x^t)+c_t(x^t)} - 0.6$$
$$\leq \frac{0.2+a_t(x^t)}{0.2+a_t(x^t)} - 0.6 = 0.4 - \frac{0.1}{0.3+a_t(x^t)}. \qquad (\star\star)$$

But, from the first recursion of (20*a*), and using the bound ($\star$) for $t+1$, we obtain, for $t \leq T-3$,

$$\frac{0.2}{0.2+m_t(x^t)} - 0.2 = m_{t+1}(x^t,1) := a_{t+1}(x^t,1) + c_{t+1}(x^t,1) \leq 0.3$$
$$\iff \quad a_t(x^t) + c_t(x^t) =: m_t(x^t) \geq 0.2$$
$$\Longrightarrow \quad a_t(x^t) \geq 0.1 \quad (\text{since} \quad c_t(x^t) < 0.1)$$

and using this bound in ($\star\star$) we obtain

$$n_{t+1}(x^t,1) \geq 0.15, \quad t \leq T-3,$$

which constitutes a contradiction to ($\bullet$), provided that $T \geq 4$.

The implication, then, is the following (negative) result.

Proposition 4.2. *For the zero-sum dynamic game of this section, and with $T \geq 4$, every saddle-point solution (p_1^*, p_2^*) has the property that, under p_2^*, the random variable s does not admit a one-point probability mass function.* ◇

The only case *Propositions 4.1* and *4.2* together do not cover is the problem with $T = 3$, which we can show for almost all parameter values also to belong to *Proposition 4.2*; but we have not included it here since a clean proof as above is not yet on paper. The message though is that for $T \geq 3$ *Player 2* has to 'mix' between different time instants, which makes the explicit computation of the saddle-point strategies quite a formidable task, especially for large values of T. An interesting open question here is whether the support set of p_2^* has to grow with T. Also, whether an approach similar to that that led to *Proposition 4.1* exists for the general case ($T \geq 3$, $S \geq 1$) is at the present an open question which requires further investigation.

5. Games with Noisy Measurements

We now turn our attention toward "noisy" versions of the game studied in the previous section, where *Player 2* does not have perfect access to *Player 1*'s positions. We do not provide a complete analysis here, but rather present some sample results and observations, which however should provide considerable insight into the intricacies of derivation of saddle points (and some of their features) for the general "noisy" version. To keep the analysis at a tractable level, we take $T = 1$ (and hence $s = 1$). Then, the cost function to be used is

$$L(u;\, v) := \begin{cases} 1 & \text{if} \quad v = x_3 \\ 0 & \text{otherwise} \end{cases} \qquad (23)$$

where $u := u_1 + u_2$, $u_i \in \{-1, 0, 1\}$, $x_3 = x_1 + u_1 + u_2$. Note that the only relevant variable for *Player 1* in this game is u, which belongs to the set $\{0, \pm 1, \pm 2\}$.

In the deterministic version of the problem treated in *section 4*, *Player 2* had perfect access to the initial position (state) x_1, which was his only measurement (since $T = 1$). Now, "noise" could be introduced into this formulation by taking the initial state to be a random quantity, on which *Player 2* could either have no additional information or make an additional imperfect (noisy) measurement. Depending on whether *Player 2* makes this additional measurement or not, we consider below two separate formulations. In both cases, *Player 1* has perfect access to the true value of x_1, and the statistical information on the random variables is *common knowledge* to both players.

"Noisy" Game 1: *Only prior statistical information for Player 2.*

The initial state x_1 is a random variable, taking two values, 0 and 1, with respective probabilities ϵ and $1 - \epsilon$, where $0 \leq \epsilon \leq 1/2$. *Player 2* does not collect any additional information on x_1, and hence his strategy space coincides with the relevant part of his action space, *i.e.* $\{0, \pm 1, \pm 2\}$, which we again denote by V, with elements v. *Player 2*, on the other hand, has access to the realized value of x_1, and hence his typical policy is in the form

$$\gamma(x_1) := \begin{cases} u^0 & \text{if } x_1 = 0 \\ u^1 & \text{if } x_1 = 1 \end{cases} \tag{24}$$

where $u^i \in U := \{0, \pm 1, \pm 2\}$, $i = 0, 1$. By a possible abuse of notation, let us denote the class of all these policies for *Player 1* again by $\mathbf{\Gamma}$. As earlier, let us introduce the sets $\mathbf{P}_1$ and $\mathbf{P}_2$, which are the collection of all probability measures on $\mathbf{\Gamma}$ and V, respectively. Elements of $\mathbf{P}_1$ and $\mathbf{P}_2$ can also be viewed as the *mixed strategies* for *Players 1* and *2*; furthermore, for *Player 1*, a mixed strategy can also be represented by a pair of random variables $r := (r_0, r_1)$, where r_0 and r_1 are both defined on V, and r_i corresponds to the choice for u at the *information node* corresponding to the realization $\{x_1 = i\}$, $i = 0, 1$. Now, under the further restriction that r_0 and r_1 are independent, a mixed strategy becomes a *behavioral strategy*; the collection of all such strategies generates a proper subset of $\mathbf{P}_1$ (to be denoted $\mathbf{P}_1^{beh}$ – the class of all behavioral strategies for *Player 1*). Since the underlying game is finite and is (trivially) of *perfect recall* (*Kuhn, 1953*), a saddle-point solution will exist even if *Player 1* is restricted to this smaller set $\mathbf{P}_1^{beh}$. The complete class of behavioral saddle-point strategies[4] for the game is given in the following Proposition.

Proposition 5.1. *The behavioral saddle-point solution for the stochastic game defined above (in terms of the action variables (u, v)) admits different characterizations in two different regions with respect to the parameter ϵ.*

i. For $\epsilon \in [0, 1/6)$, the behavioral saddle-point strategies are **unique** *and are given by:*

$$Prob(u = i \mid x_1 = 0) = 1, \qquad i = -2 \tag{25a}$$

[4] *For Player 2, every mixed strategy is also a behavioral strategy, since he has a single information set.*

$$Prob(u = i \mid x_1 = 1) = \frac{1}{5}, \qquad i = 0, \pm 1, \pm 2 \tag{25b}$$

$$Prob(v = j) = \frac{1}{5}, \qquad j = 0, \pm 1, 2, 3 \tag{26}$$

which lead to a saddle-point value of $\frac{1}{5}(1-\epsilon)$.

ii. For $\epsilon \in [1/6, 1/2]$, *the complete class of behavioral saddle-point strategies are:*

$$Prob(u = i \mid x_1 = 0) = \begin{cases} \frac{1}{4} - \frac{1}{24\epsilon} + \alpha_i^0 & i = 0, \pm 1, 2 \\ \frac{1}{6\epsilon} & i = -2 \end{cases} \tag{27a}$$

$$Prob(u = i \mid x_1 = 1) = \begin{cases} \frac{1}{4} - \frac{1}{24(1-\epsilon)} + \alpha_{i+1}^1 & i = 0, \pm 1, -2 \\ \frac{1}{6(1-\epsilon)} & i = 2 \end{cases} \tag{27b}$$

$$Prob(v = j) = \frac{1}{6}, \qquad j = 0, \pm 1, \pm 2, 3 \tag{28}$$

where α_i^0*'s and* α_i^1*'s are arbitrary scalars satisfying the relationships:*

$$\epsilon \alpha_i^0 + (1-\epsilon)\alpha_i^1 = 0, \qquad i = 0, \pm 1, 2$$

$$\alpha_i^0 \geq \frac{1}{24\epsilon} - \frac{1}{4}; \ \alpha_i^1 \geq \frac{1}{24(1-\epsilon)} - \frac{1}{4}, \quad i = 0, \pm 1, 2$$

$$\sum_{i=-1}^{2} \alpha_i^0 = \sum_{i=-1}^{2} \alpha_i^1 = 0 .$$

The saddle-point value of the game is $\frac{1}{6}$.

Proof. We present the proof separately in the two regions.

i. For the region $\epsilon \in [0, 1/6)$, we direct our attention to establishing the uniqueness part of the result; as a by-product, we also obtain the saddle-point property of the given strategies. Toward this end, let us first compute the best response of *Player 1* to (26) when he is at the information node $\{x_1 = 0\}$. Under the measure $\mathbf{p}_2^*$ corresponding to (26),

$$E^{\mathbf{p}_2^*}[L(u; v) \mid u = i, x_1 = 0] = \begin{cases} 0 & \text{if } i = -2 \\ 1/5 & \text{if } i = 0, \pm 1, 2 \end{cases}$$

from which it readily follows that the **unique** minimizing solution for *Player 1* is the ***one-point*** measure (25*a*). Then, because of the ***ordered interchangeability*** property of multiple saddle-point equilibria (*Başar and Olsder, 1982*), (25*a*) is the only saddle-point strategy for *Player 1* at the information node $\{x_1 = 1\}$, for all ϵ in the given interval. Now, to prove the uniqueness of (25*b*), we first assume (to the contrary) that there exists an equilibrium strategy for *Player 1*, at the node $\{x_1 = 1\}$, that is different from (25*b*). Such a strategy has to be given by

$$Prob(u = i \mid x_1 = 1) = q_i\,, \quad i = 0, \pm 1, \pm 2, \qquad (\bullet)$$

where the q_i's are all nonnegative but not all equal (note that $\sum_{i=-2}^{2} q_i = 1$). Under the (behavioral strategy) measure $\mathbf{p}_1^*$ that corresponds to (25*a*) and ($\bullet$), *Player 2*'s best response is obtained by the maximization of the following function over *V*:

$$E^{\mathbf{p}_1^*}[L(u;\, v) \mid v] = \begin{cases} (1-\epsilon)q_{j-1} & v = j\,, \quad j \in \{0, \pm 1, 2, 3\} \\ \epsilon & v = -2 \end{cases} .$$

Since $0 \leq \epsilon \leq 1/6$, the point $\{v = -2\}$ can never be in the maximizing set, and for (26) to provide a solution it is necessary that the function above be a constant on the set $\{0, \pm 1, 2, 3\}$, which is only possible if all the q_i's are equal (and hence 1/5). Thus, (25*b*) provides a **unique** equilibrium strategy for *Player 1* at the information node $\{x_1 = 1\}$. Finally, to prove the uniqueness of (26), we first introduce a measure $\mathbf{p}_2$, corresponding to the choice

$$\mathbf{p}_2 : \quad Prob(v = j) = p_j\,, \quad j = 0, \pm 1, 2, 3\,,$$

where the p_j's are nonnegative and sum up to 1. Note that we have not put any weight on the point $\{v = -2\}$, since it cannot appear in any saddle-point strategy by the argument above (in the proof of uniqueness of (25*b*)). Now, under the measure $\mathbf{p}_2$, the response of *Player 1* at the node $\{x_1 = 1\}$ is obtained from the minimization of

$$E^{\mathbf{p}_2}[L(u;\, v) \mid u = i,\, x_1 = 1] = p_{i+1}\,, \qquad i \in U\,,$$

which admits (25*b*) as a minimizing solution only if all the p_i's are identical. Hence, the solution given by (26) is the **unique** choice for *Player 2* that is in equilibrium with (25). This completes the proof of the *Proposition* in the region $\epsilon \in [0, 1/6]$.

ii. For the region $\epsilon \in [1/6, 1/2]$, it is again fairly straightforward to show that the given set of policies constitute a saddle point in behavioral strategies; what is somewhat more involved, however, is the verification of the completeness of this class. We therefore focus on that point, and first show that (28) is a unique behavioral saddle-point policy for *Player 2*. Suppose the contrary is true; *i.e.* there are nonnegative scalars, p_j's, summing up to zero, and not all identical, such that the measure $\mathbf{p}_2$ given by

$$\mathbf{p}_2 : \quad Prob(v = j) = p_j\,, \quad j = 0, \pm 1, \pm 2, 3\,,$$

provides another behavioral saddle-point strategy. We now compute

$$E^{\mathbf{p}_2}[L(u;\, v) \mid u = i,\, x_1 = 0] = p_i\,, \qquad i = 0, \pm 1, \pm 2\,,$$

and observe that for (27*a*) (with, for example, all the α_i^0's taken to be zero) to be a minimizing solution for this function, all p_i's, with a possible exception of p_3, have to be identical, so that the expression is independent of *u* on *U*. To prove that p_3 has to have the same value, we now consider the counterpart of the above expression for $x_1 = 1$:

$$E^{\mathbf{p}_2}[L(u;v)\mid u=i,\, x_1=1]=p_{i+1}\,, \qquad i=0,\pm1,\pm2\,,$$

from which it readily follows (by a similar argument) that p_3 cannot be different in value from the others. Now, to prove the completeness of (27), we replace (27*a*) with the set of nonnegative scalars $\{q_i^0,\ i\in U\}$, and (27*b*) with a similar set $\{q_i^0,\ i\in U\}$, and under this measure (to be denoted $\mathbf{p}_1^q$) we determine the maximization problem faced by *Player 2*:

$$E^{\mathbf{p}_1^q}[L(u;v)\mid v=j]=\begin{cases}\epsilon q_j^0+(1-\epsilon)q_{j-1}^1 & j\in\{0,\pm1,2\}\\ (1-\epsilon)q_2^1 & j=3\\ \epsilon q_{-2}^0 & j=-2\end{cases}$$

Since (28) assigns nonzero (in fact, equal) weight to all elements in the set $\{0,\pm1,\pm2,3\}$, the above expression has to have the constant value (1/6) on that set. Noting also that the q_i^0's and q_i^1's are probability weights (and therefore each should sum up to 1), a simple algebra leads to the conclusion that the given set (27) is complete. ◇

Several observations are now in order. The first is that for a small level of uncertainty in the value of the initial state (precisely, for $\epsilon<1/6$), the maximizer's saddle-point strategy is insensitive to the presence of this uncertainty – hence, the policy for the deterministic problem (with $x_1=1$) is fairly *robust.* The second observation is that for large values of uncertainty (specifically, for $1/6<\epsilon\le 1/2$) the maximer's saddle-point policy is the same as the one that ignores the prior information completely, and the saddle-point value is a constant, independent of the precise level of uncertainty. Hence, the unique saddle-point policy of *Player 2* features a threshold property with respect to the level of uncertainty. *Player 1*'s saddle-point strategies, on the other hand, depend critically on both his information and the level of uncertainty (faced by the other player) on the value of the initial state.

"Noisy" Game 2: *Measurement information for Player 2.*

Here we assume that *Player 2* makes some noisy measurement of the random initial state, with the observation variable being y, taking the two possible values 0 and 1. To permit an interesting comparison between the game solved above and the one to be analyzed here, we take the conditional probabilities $Prob(y=i\mid x_1=j)$ to be equal to ϵ for $i\neq j$, and $1-\epsilon$ for $i=j,\ i,j=0,1$ where $0<\epsilon<1/2$. We further assume that the prior statistics on x_1 provides no additional information, in that $Prob(x_1=i)=1/2,\ i=0,1$. The complete solution to this game is now provided below.

Proposition 5.2. *The behavioral saddle-point solution for the stochastic "noisy" game 2 defined above is uniquely given by the following probability mass functions:*

$$Prob(u=i\mid x_1=0)=\begin{cases}\dfrac{1}{5-4\epsilon} & i=-2,\\[2mm] \dfrac{1-\epsilon}{5-4\epsilon} & i=0,\pm1,2\end{cases} \tag{29a}$$

$$Prob(u = i \mid x_1 = 1) = \begin{cases} \dfrac{1-\epsilon}{5-4\epsilon} & i = -2, 0, \pm 1 \\ \dfrac{1}{5-4\epsilon} & i = 2 \end{cases} \tag{29b}$$

$$Prob(v = j \mid y = 0) = \begin{cases} \dfrac{1}{5-4\epsilon} & j = -2, \\ \dfrac{1-\epsilon}{5-4\epsilon} & j = 0, \pm 1, 2 \end{cases} \tag{30a}$$

$$Prob(v = j \mid y = 1) = \begin{cases} \dfrac{1-\epsilon}{5-4\epsilon} & j = 0, \pm 1, 2 \\ \dfrac{1}{5-4\epsilon} & j = 3 \end{cases} \tag{30b}$$

The saddle-point value of the game is $(1-\epsilon)/(5-4\epsilon)$.

Proof. The fact that (29) and (30) are in equilibrium follows from a routine verification of the corresponding pair of saddle-point inequalities. We therefore concentrate our analysis here on proving the completeness (in this case uniqueness) of the policies (29)-(30). Toward this end, let

$$Prob(u = i \mid x_1 = 0) = p_i \; ; \quad Prob(u = i \mid x_1 = 1) = q_i \, , \quad i = 0, \pm 1, \pm 2 \, ,$$

where p_i's and q_i's are arbitrary probability weights attached to the different alternatives. Denote such a typical probability measure for *Player 1* by $\mathbf{p}_1$. Then, under $\mathbf{p}_1$, and at the information node $\{y = 0\}$, the optimal response policies for *Player 2* are obtained by the maximization of the quantity $E^{\mathbf{p}_1}[L(u;\, v) \mid y = 0]$ over $v \in \mathbf{Z}$, where

$$E^{\mathbf{p}_1}[L(u;\, v) \mid y = 0,\, v = j] = \begin{cases} (1-\epsilon)p_j + \epsilon q_{j-1} \, , & j \in \{0, \pm 1, 2\} \\ \epsilon q_2 \, , & j = 3 \\ (1-\epsilon)p_2 \, , & j = -2 \end{cases} .$$

In view of (30), the support set of the maximizing solution should include the set $\{0, \pm 1, \pm 2\}$, and hence the following relations should hold:

$$\begin{aligned} (1-\epsilon)p_{-2} &= (1-\epsilon)p_{-1} + \epsilon q_{-2} = (1-\epsilon)p_0 + \epsilon q_{-1} = (1-\epsilon)p_1 + \epsilon q_0 \\ &= (1-\epsilon)p_2 + \epsilon q_1 \quad \geq \epsilon q_2 \, . \end{aligned} \tag{i}$$

Likewise, at the information node $\{y = 1\}$, *Player 2* is faced with the maximand

$$E^{\mathbf{p}_1}[L(u;\, v) \mid y = 1,\, v = j] = \begin{cases} \epsilon p_j + (1-\epsilon)q_{j-1} \, , & j \in \{0, \pm 1, 2\} \\ (1-\epsilon)q_2 \, , & j = 3 \\ \epsilon p_2 \, , & j = -2 \end{cases} ,$$

and by the same argument as before, this time using the set $\{0, \pm 1, 2, 3\}$ in view of (30b), we arrive at the relationships

$$(1-\epsilon)q_2 = \epsilon p_{-1} + (1-\epsilon)q_{-2} = \epsilon p_0 + (1-\epsilon)q_{-1} = \epsilon p_1 + (1-\epsilon)q_0$$
$$= \epsilon p - 2 + (1-\epsilon)q_1 \quad \geq \epsilon p_{-2}\,. \qquad (ii)$$

It follows through some routine algebra that, under the constraints $\sum_{i=-2}^{2} p_i = \sum_{i=-2}^{2} q_i = 1$, (i) and (ii) admit the **unique** solution

$$q_2 = p_{-2} = \frac{1}{5-4\epsilon}\,; \quad p_i = q_{i-1} = \frac{1-\epsilon}{5-4\epsilon}\,,\ i = 0, \pm 1, 2,$$

and the inequalities in both (i) and (ii) are strict since $\epsilon<1/2$. Note that this latter property implies that for any saddle-point solution for *Player 2*,

$$Prob(v = -2 \mid y = 1) = Prob(v = 3 \mid y = 0) = 0\,.$$

Now, to prove the uniqueness of (30), we can take, without loss of generality,

$$Prob(v = j \mid y = 0) = p_j\,; \quad Prob(v = j \mid y = 1) = q_j\,, \quad j = 0, \pm 1, \pm 2, 3\,,$$

with $p_3 = q_{-2} = 0$, and where p_j's and q_j's are again probability weights, separately summing up to 1. Let the measure corresponding to such a strategy be $\mathbf{p}_2$. Then, we have the following functions to be minimized by *Player 1*:

$$E^{\mathbf{p}_2}[L(u;v) \mid x_1 = 0, u = j] = \begin{cases} (1-\epsilon)p_j + \epsilon q_j\,, & j \in \{0, \pm 1, 2\} \\ (1-\epsilon)p_{-2}\,, & j = -2 \end{cases}\,,$$

$$E^{\mathbf{p}_2}[L(u;v) \mid x_1 = 1, u = j] = \begin{cases} \epsilon p_{j+1} + (1-\epsilon)q_{j+1}\,, & j \in \{0, \pm 1, -2\} \\ (1-\epsilon)q_3\,, & j = 2 \end{cases}\,.$$

Since the strategy (29) is totally mixed, the above functions should be constants over U, implying that

$$(1-\epsilon)p_{-2} = (1-\epsilon)p_j + \epsilon q_j\,, \quad j = 0, \pm 1, 2 \qquad (iii)$$

$$(1-\epsilon)q_3 = \epsilon p_{j+1} + (1-\epsilon)q_{j+1}\,, \quad j = 0, \pm 1, -2\ . \qquad (iv)$$

Again one can show that the set of equations (iii) and (iv), together with the basic equation $\sum_{j=-2}^{2} p_j = \sum_{j=-2}^{2} q_{j+1} = 1$, admits the **unique** solution

$$p_{-2} = q_3 = \frac{1}{5-4\epsilon}\,; \quad p_j = q_{j+1} = \frac{1-\epsilon}{5-4\epsilon}\,,\ j = 0, \pm 1, 2.$$

This completes the proof of the Proposition. ◇

We are now in a position to compare the two solutions presented in *Proposition 5.1* and *Proposition 5.2*. Note that if in **game 2** *Player 2* makes the observation $y = 1$, then the uncertainty he faces as to the true value of x_1 is the same as in **game 1**. However, with probability 1/2 *Player*

2 could also make the measurement $y = 0$, and the actual realization (*i.e.* whether y=0 or y=1 has been observed by *Player 2*) is not known to *Player 1*. This actually puts *Player 1* at a disadvantage in **game 2** (as compared with **game 1**), as a result of which the saddle-point value in **game 1** becomes strictly dominating (except at $\epsilon = 0$) over that of **game 2**, to the advantage of *Player 1* – the minimizer. Yet another striking difference between the two solutions is that for small levels of uncertainty (specifically, for ϵ in the range $0 < \epsilon < 1/6$), the unique saddle-point policies for the players in **game 1** are independent of the precise value of the quantification of the uncertainty (*i.e.*, ϵ), whereas in **game 2** they show explicit dependence on ϵ. As mentioned before, this indicates some *robustness* of the saddle-point policies in **game 1** to small levels of uncertainty, which is indeed a desirable feature. This extends also to the case where the action set U is much larger (as in *Section 3*), under a similar description of uncertainty and with *Player 2* not making any measurements. For the stochastic version of the more general game of *Section 4*, however, where *Player 2* acts more than once, under "noisy" observations and with memory, such a feature may no longer hold since a multi-stage game will inevitably involve *learning* on the part of the player making the noisy measurements. Derivation of the saddle-point solution for such a multi-stage dynamic stochastic game seems to be quite a challenging, but promising, task for the future. A further extension would be to the class of stochastic discrete games defined on the plane, where (as discussed briefly in *section 3*) the search is now on a two-dimensional grid.

Acknowledgement. I would like to thank Dr. Pierre Bernhard (INRIA, Sophia Antipolis) for various discussions on the topic of this paper, which have been the motivating force behind the research that led to the results reported here. Further thanks go to Dr. V.J. Baston and Dr. F.A.B. Bostock (University of Southampton) for bringing to my attention an error in an earlier draft of the paper, which has led to a reformulation of the problem and to the current section 4.

References

Başar, T. and G. J. Olsder (1982), *Dynamic Noncooperative Game Theory*, Academic Press, London/New York.

Bernhard, P. (1987), "Computation of equilibrium points of delayed partial information pursuit evasion games", *Proceedings of the 26th IEEE Conference on Decision and Control*, Los Angeles, CA, December 9-11, pp. 281-283.

Bernhard, P. and A.-L. Colomb (1988), "Saddle point conditions for a class of stochastic dynamical games with imperfect information", *IEEE Transactions on Automatic Control*, vol. AC-33, no. 1, pp. 98-101.

Bernhard, P., A.-L. Colomb and G. Papavassilopoulos (1987), "Rabbit and Hunter game: two discrete stochastic formulations", *Comput. Math. Applic.*, vol. 13, no. 1-3, pp. 205-225.

Ferguson, T. S. (1967), "On discrete evasion games with a two-move information lag", *Proc. 5th Berkeley Symp. Math. Stat. Probability*, vol. 1, pp. 453-462.

Kuhn, H. W. (1953), "Extensive games and the problem of information", in *Contributions to the Theory of Games*, H.W.Kuhn and A.W.Tucker, eds; pp. 193-216; Annals of Mathematics Studies, no. 28, Princeton University Press, Princeton, New Jersey.

A SURVEY OF RECENT RESULTS IN DIFFERENTIAL GAMES*

LEONARD D. BERKOVITZ
Purdue University, West Lafayette, IN 47907, USA

1. Introduction

In this paper we summarize some of our recent results on zero–sum two person differential games. Formulated intuitively, a differential game has its state at time t given by a vector $x(t)$ in R^n determined by a system of differential equations

$$\frac{dx}{dt} = f(t, x, u(t), v(t)), \quad x(t_0) = x_0. \tag{1.1}$$

where $u(t)$ is chosen by Player I at each instant of time and $v(t)$ is chosen by Player II at each instant of time. The choices are constrained by the conditions $u(t) \in Y$ and $v(t) \in Z$, where Y and Z are preassigned sets in euclidean spaces. The choice of $u(t)$ is governed by a "strategy" U selected by Player I prior to the start of play and the choice of $v(t)$ is governed by a "strategy" V selected by Player II prior to the start of play. Play proceeds from the initial point (t_0, x_0) until the point $(t, \phi(t))$, where ϕ is the solution of (1.1), reaches some preassigned terminal set $\mathfrak{T}$. The point at which $(t, \phi(t))$ reaches $\mathfrak{T}$ is called the terminal point and is denoted by $(t_f, \phi(t_f))$ or (t_f, x_f). The payoff is

$$P(t_0, x_0;\ U, V) = g(t_f, x_f) + \int_{t_0}^{t_f} f(s, \phi(s), u(s), v(s))ds. \tag{1.2}$$

Player I wishes to choose U so as to maximize P while Player II wishes to choose V so as to minimize P.

If

$$\sup_U \inf_V P(t_0, x_0;\ U, V) = \inf_V \sup_U (t_0, x_0;\ U, V)$$

*This paper was prepared under National Science Foundation Grant No. DMS 8700813

then we denote this number by $W(t_0, x_0)$ and say that the game has value equal to $W(t_0, x_0)$. A pair of strategies (U^*, V^*) is said to be a saddle point if for all U over which the sup is taken and all V over which the inf is taken

$$P(t_0, x_0;\ U, V^*) \leq P(t_0, x_0;\ U^*, V^*) \leq P(t_0, x_0;\ U^*, V).$$

The strategies U^* and V^* are called optimal strategies.

The first problem associated with differential games is to formulate the intuitive notion as a game in the mathematical sense of a zero–sum two person game. This involves defining the notion of strategy or strategy space precisely and defining the corresponding notion of payoff as a function on the strategy spaces of the players. The second problem is to determine whether for the given definition of strategy the game has a value. The third problem is to determine whether the game has a saddle point in the event that it has a value. The fourth problem is to find the optimal strategies in the event that a saddle point exists.

If one perceives the game as being a game of perfect information one might want to take the strategies U and V to be functions of (t, x) with appropriate properties and substitute these functions into (1.1) for $u(t)$ and $v(t)$ respectively. This approach, however, has had only limited success. For a discussion of the drawbacks and the successes, see [1]. In the present work, the direct approach is abandoned. Instead, indirect definitions are used which first involve choices of control functions u, v on small intervals, with the choices being made at discrete times and based on the history of the game. A passage to the limit as the lengths of the small intervals tend to zero is then carried out.

2. Assumptions and Notation

Let t denote time, let $x = (x^1, \cdots, x^n)$ denote a vector in n-dimensional real euclidean space R^n, let $y = (y^1, \cdots, y^r)$ denote a vector in R^r and let $z = (z^1, \cdots, z^s)$ denote a vector in R^s. We use superscripts to denote components of a vector and use subscripts to distinguish vectors. We shall denote the inner product of two vectors x and y by $\langle x, y \rangle$, and the euclidean norm of a vector x by $|x|$.

Let Y be a compact set in R^r, let Z be a compact set in R^s and let T_0 and T_1 be two real numbers satisfying $T_0 < T_1$. Let $f^0, f^1, \cdots, f^n$ be real valued functions defined on the set

$$\mathcal{D} = [T_0, T_1] \times R^n \times Y \times Z \tag{2.1}$$

and let $f = (f^1, \cdots, f^n)$. Let $\hat{f} = (f^0, f)$. Let g be a real valued function defined on $[T_0, T_1] \times R^n$ and bounded on bounded subsets of its domain. Let $\mathcal{T}$ be a closed set in (t, x)–space such that

$$[T_0, \infty) \times R^n \supset \mathcal{T} \supseteq [T_1, \infty) \times R^n. \tag{2.2}$$

The function f will govern the dynamics as suggested by (1.1). The functions f^0 and g will enter the payoff as in (1.2) and the set $\mathcal{T}$ will be the terminal set. The precise interpretation of (1.2) will be given in Section 3.

Assumption I

(i) The function $\hat{f}$ is continuous on $\mathcal{D}$.

(ii) There exists a constant $K > 0$ such that for all $t \in [T_0 T_1]$, $x, \bar{x}$ in R^n, $y \in Y$ and $z \in Z$

$$|f(t, x, y, z) - f(t, \bar{x}, y, z)| \leq K|x - \bar{x}|.$$

(iii) There exists a constant $K' > 0$ such that for all $t, \bar{t}$ in $[T_0, T_1]$, $x, \bar{x}$ in R^n,

$$|g(\bar{t}, \bar{x}) - g(t, x)| \leq K'(|t - \bar{t}| + |x - \bar{x}|).$$

3. Definition of Game

Our definition of strategy is related to the "K strategies" of Isaacs [11] and the lower strategies of Friedman [10]. Our definition of payoff is related to that of Krasovskii–Subbotin [13].

For any interval $[a, b] \subseteq [T_0, T_1]$ let $\mathcal{Y}[a, b]$ denote the set of measurable functions u on $[a, b]$ such that $u(t) \in Y$ a.e. Let $\mathcal{Z}[a, b]$ denote the set of measurable functions v on $[a, b]$

such that $v(t) \in Z$ a.e. Functions u in $\mathcal{Y}[a,b]$ will be called controls for Player I on $[a,b]$ and functions v in $\mathcal{Z}[a,b]$ will be called controls for Player II on $[a,b]$.

A strategy for Player I for the game starting at (t_0, x_0) is a choice of a sequence of partitions $\Pi = \{\Pi_n\}$ of $[t_0, T_1]$ and a choice of a sequence of maps $\Gamma_\Pi = \{\Gamma_{\Pi,n}\}$, where the $\Gamma_{\Pi,n}$ are to be defined below. Thus $\Gamma = \{\Gamma_\Pi, \Pi\}$. For typographic simplicity, we shall suppress the dependence on Π in the notation and shall write Γ for Γ_Π and $\{\Gamma_n\}$ for $\{\Gamma_{\Pi,n}\}$. We restrict the choice of sequences of partitions to those such that $\|\Pi_n\| \to 0$ as $n \to \infty$. Let the partition points of Π_n be $t_0 < t_1 < \cdots < t_p = T_1$. Each Γ_n is a collection of maps $\Gamma_{n,1}, \cdots, \Gamma_{n,p}$ as follows. The map $\Gamma_{n,1}$ selects an element in $\mathcal{Y}[t_0, t_1)$. For $2 \le j \le p$ the map $\Gamma_{n,j}$ is a map from $\mathcal{Y}[t_0, t_{j-1}) \times \mathcal{Z}[t_0, t_{j-1})$ to $\mathcal{Y}[t_{j-1}, t_j)$.

A strategy for Player II is a choice of a sequence of partitions $\overline{\Pi} = \{\overline{\Pi}_n\}$ of $[t_0, T_1]$ such that $\|\overline{\Pi}_n\| \to 0$ as $n \to \infty$ and a choice of a sequence of maps $\{\Delta_n\}$. If $\overline{\Pi}_n$ has partition points $t_0 = s_0 < s_1 < \cdots < s_q = T_1$, then each Δ_n is a collection of maps $\Delta_{n,1}, \Delta_{n,2}, \cdots, \Delta_{n,q}$ as follows. The map $\Delta_{n,1}$ selects a function v in $\mathcal{Z}[s_0, s_1)$. For $2 \le j \le q$, $\Delta_{n,j}$ is a map from $\mathcal{Y}[t_0, s_{j-1}) \times \mathcal{Z}[t_0, s_{j-1})$ to $\mathcal{Z}[s_{j-1}, s_j)$. The collection of maps $\Gamma_n = (\Gamma_{n,1}, \cdots, \Gamma_{n,p})$ will be called an n-th stage strategy for Player I. The collection of maps $\Delta_n = (\Delta_{n,1}, \cdots, \Delta_{n,q})$ will be called an n-th stage strategy for Player II.

We next define the concept of a motion in the game. If w is a measurable function defined on $[t_0, T_1]$ and if $a \in (t_0, T_1)$ by $w|a$ we shall mean the restriction of w to $[t_0, a)$. A pair of n-th stage strategies (Γ_n, Δ_n) determine a control u_n on $[t_0, T_1]$ for Player I and a control v_n for Player II on $[t_0, T_1]$ as follows.

$$
\begin{aligned}
u_n(t) &= (\Gamma_{n,1})(t) \qquad \text{for} \quad t \in [t_0, t_1) \\
v_n(t) &= (\Delta_{n,1})(t) \qquad \text{for} \quad t \in [t_0, s_1).
\end{aligned}
\tag{3.1}
$$

By considering the various possibilities that may occur in the natural ordering of the union of the partition points $\{t_i\}_{i=0}^p$ and $\{s_i\}_{i=0}^q$, the reader will convince himself that the functions u_n and v_n are indeed well defined on $[t_0, T_1]$ by formula (3.1) and formula (3.2) which follows.

$$
\begin{aligned}
u_n(t) &= \Gamma_{n,j}(u_n|t_{j-1}, v_n|t_{j-1})(t) \qquad t \in [t_{j-1}, t_j) \qquad 2 \le j \le p \\
v_n(t) &= \Delta_{n,j}(u_n|s_{j-1}, v_n|s_{j-1})(t) \qquad t \in [s_{j-1}, s_j) \qquad 1 \le j \le q.
\end{aligned}
\tag{3.2}
$$

The controls (u_n, v_n) determined this way are called the n–th stage outcomes of (Γ_n, Δ_n).

In equation (1.1), if we replace $u(t)$ by $u_n(t), v(t)$ by $v_n(t)$ and x_0 by x_{0n} we obtain the system of differential equations

$$\frac{dx}{dt} = f(t, x, u_n(t), v_n(t)) \qquad x(t_0) = x_{0n}. \tag{3.3}$$

It follows from Assumption I that equation (3.3) has a unique solution $\phi_n(\ , t_0, x_{0n}, u_n, v_n)$ defined on $[t_0, T_1]$. We call this solution an n–th stage trajectory.

Any uniform limit of a subsequence of the n–th stage trajectories $\phi_n(\ , t_0, x_{0n}, u_n, v_n)$, $n = 1, 2, 3, \cdots$ where $x_{0n} \to x_0$ and (u_n, v_n) is the outcome of (Γ_n, Δ_n) will be called a *motion* or *motion of the game* corresponding to strategies $\Gamma = \{\Gamma_n\}$ and $\Delta = \{\Gamma_n\}$. Following Krasovskii and Subbotin [13] we denote a motion corresponding to (Γ, Δ) by $\phi[\ , t_0, x_0, \Gamma, \Delta]$.

It follows from Assumption I that corresponding to a pair of strategies (Γ, Δ) and a sequence of initial conditions $\{x_{0n}\}$ with $x_{0n} \to x_0$, the corresponding sequence of n–th stage trajectories $\{\phi_n(\ , t_0, x_{0n}, u_n, v_n)\}$ is uniformly bounded and equicontinuous. Hence motions $\phi[\ , t_0, x_1, \Gamma, \Delta]$ do exist. We shall denote the set of all motions corresponding to (Γ, Δ) by $\Phi[\ , t_0, x_0, \Gamma, \Delta]$.

By $\{\phi[t, t_0, x_{0n}, u_n, v_n)\}$ we shall mean the motion evaluated at the point t. By $\Phi[t, t_0, x_0, \Gamma, \Delta]$ we shall mean the set of all values $\phi[t, t_0, x_0, \Gamma, \Delta]$, where $\phi[\ , t_0, x_0, \Gamma, \Delta]$ ranges over $\Phi[\ , t_0, x_0, \Gamma, \Delta]$.

It is important to keep in mind that, in general, a motion $\phi[\ , t_0, x_0, \Gamma, \Delta]$ will not be obtained as a solution of

$$\frac{dx}{dt} = f(t, x, u(t), v(t)) \qquad x(t_0) = x_0$$

for an appropriate choice of controls u, v. Nevertheless, motions do have properties similar to or analogous to properties of solutions of differential equations. For example, the continuous dependence on initial data has the following analogue, imprecisely stated. If (τ', ξ') is near (τ, ξ) then for each motion $\phi[\ , \tau, \xi, \Gamma, \Delta]$ there exists a motion $\phi[\ , \tau', \xi'\Gamma', \Delta']$ that is close to $\phi[\ ,\tau,\xi,\Gamma,\Delta]$ in the sup norm. For the precise statement and proof of this property and others, see [2].

To complete the definition of the game we must define the payoff. We first consider games of fixed duration. In such games the terminal set $\mathcal{T}$ is given by

$$\mathcal{T} = [T, \infty) \times R^n,$$

where $T_0 < T \leq T_1$. The function g is a mapping $x \to g(x)$ and is defined on R^n. If $f^\circ \equiv 0$ the game is said to have terminal payoff. There is no loss of generality in assuming $f^\circ \equiv 0$. To see this let $\hat{x} = (x^0, x)$ and adjoin the equation

$$\frac{dx^0}{dt} = f^0(t, x, u(t), v(t)) \qquad x^0(t_0) = 0 \tag{3.4}$$

to the system (1.1). The game with payoff (1.2) becomes a game of fixed duration in $[T_0, T_1] \times R^{n+1}$ with terminal payoff $\hat{g}(\hat{x}) = g(x) + x^0$.

For games of fixed duration with terminal payoff g the payoff corresponding to a pair of strategies (Γ, Δ) is set valued and is defined by

$$P(t_0, x_0, \Gamma, \Delta) = g(\Phi[T, t_0, x_0, \Gamma, \Delta])$$

We next consider games with $\mathcal{T}$ as in (2.2) and $g \equiv 0$ in (1.2). Such games are often called games of generalized pursuit and evasion. We treat such games by adjoining (3.4) to (1.1) as in the preceding paragraph and study the resulting system in R^{n+1}. The sequence of n-th stage trajectories corresponding to a pair of strategies (Γ, Δ) is the sequence $\{\hat{\varphi}_n(\ , t_0, x_{0n}, u_n, v_n)\}$, the solutions of

$$\frac{d\hat{x}}{dt} = \hat{f}(t, x, u_n(t), v_n(t)) \qquad \hat{x}(t_0) = \hat{x}_{0n}, \tag{3.5}$$

where $\hat{x}_n \to (0, x_0)$. By the capture time t_f of a motion $\hat{\varphi}[\ , t_0, x_0, \Gamma, \Delta]$ we mean the first time t_f such that $(t, \varphi[t, t_0, x_0, \Gamma, \Delta]) \in \mathcal{T}$. The payoff is now defined by

$$P(t_0, x_0, \Gamma, \Delta) = \cup\varphi^0[t_f, t_0, x_0, \Gamma, \Delta],$$

where the union is taken over all motions $\hat{\varphi}[\ ,t_0,x_0,\Gamma,\Delta]$ resulting from (Γ,Δ). Note that in general t_f will be different for different motions.

Lastly we consider the game with the terminal set as in (2.2) and the payoff as in (1.2). Such games are often called games of survival. To define the payoff we again take the sequence of n–th stage trajectories coresponding to a pair of strategies (Γ,Δ) to be the sequence $\{\hat{\varphi}_n(\ ,t_0,x_{0n},u_n,v_n)\}$, where $\hat{\varphi}_n$ is the solution of (3.5). The capture time t_f of a motion is as in the preceding paragraph. The payoff $P(t_0,x_0,\Gamma,\Delta)$ resulting from a pair of strateiges (Γ,Δ) is defined as

$$P(t_0,x_0,\Gamma,\Delta) \equiv \cup\left\{g(\varphi[t_f,t_0,x_0,\Gamma,\Delta]) + \varphi^0[t_f,t_0,x_0,\Gamma,\Delta]\right\},$$

where the union is taken over all motions $\hat{\varphi}[\ ,t_0,x_0,\Gamma,\Delta]$ resulting from (Γ,Δ).

We conclude this section by presenting two other definitions of strategies. Let $\mathcal{Y}_c[a,b]$ denote the set of piecewise constant functions u on $[a,b]$ with values $u(t)$ in Y and let $\mathcal{Z}_c[a,b]$ denote the set of piecewise constant functions v on $[a,b]$ with values $v(t)$ in Z. We define strategies as before except that now $\Gamma_{n,1}$ selects a vector $y_{n,1}$ in Y and $u_n(t) = y_{n,1}$ for all $t \in [t_0,t_1)$. Similar $\Delta_{n,1}$ selects a vector $z_{n,1}$ in Z and $v_n(t) = z_{n,1}$ for all $t \in [t_0,s_1)$. For $2 \le j \le p$, $\Gamma_{n,j}$ is a map from $\mathcal{Y}_c[t_0,t_{j-1}) \times \mathcal{Z}_c[t_0,t_{j-1})$ to Y and for $t \in [t_{j-1},t_j)$, $u_n(t) = (\Gamma(u_n|t_{j-1},v_n|t_{j-1}))$. For $2 \le j \le q$, $\Delta_{n,j}$ is a map from $\mathcal{Y}_c[t_0,s_{j-1}) \times \mathcal{Z}_c[t_0,s_{j-1})$ to Z and for $t \in [s_{j-1},s_j)$, $v_n(t) = (\Delta(u_n|s_{j-1},v_n|s_{j-1}))$. Thus, loosely speaking, at a decision point t_j or s_j an n–th stage strategy selects a constant valued control over the succeeding interval. Motions are now defined as before, except that we use trajectories obtained from these (u_n,v_n). Also defined as before are the various payoffs. We shall denote the game so defined as G_c.

To define the second version, we let $\mathcal{Y}_r[a,b]$ denote the set of relaxed controls on $[a,b]$ relative to the compact set Y. We let $\mathcal{Z}_r[a,b]$ denote the set of relaxed controls on $[a,b]$ relative to the compact set Z. For a summary of properties of relaxed controls see [2],[4]. For a full discussion see [14]. We define a strategy for Player I as before except that now $\Gamma_{n,1}$ selects a relaxed control on $[t_0,t_1)$ and $\Delta_{n,1}$ selects a relaxed control on $[t_0,s_1)$. For $2 \le j \le p$, $\Gamma_{n,j}$ is a map from $\mathcal{Y}_r[t_0,t_{j-1}) \times \mathcal{Z}_r[t_0,t_{j-1})$ to $\mathcal{Y}_r[t_{j-1},t_j)$. Similarly, for

$2 \leq j \leq q$, $\Delta_{n,j}$ is a map from $\mathcal{Y}_r[t_0, s_{j-1}) \times \mathcal{Z}_r[t_0, s_{j-1})$ to $\mathcal{Z}_r[s_{j-1}s_j)$. A pair of n–th stage strategies (Γ_n, Δ_n) determined a pair of n–th stage relaxed controls (η_n, ζ_n). These relaxed controls determine a sequence of n–th stage relaxed trajectories via the differential equation

$$\frac{dx}{dt} = f(t, x, \eta_n(t), \zeta_n(t)) \equiv \int_Z \int_Y f(t, x, y, z) d\eta_n(t) d\zeta_n(t).$$

Motions and payoffs are defined as before using relaxed trajectories and strategies as defined here. For details see [4]. We shall call the game just described the game G_r.

4. The Upper and Lower Values, Saddle Points

Let

$$\begin{aligned} W^-(t_0, x_0) &= \sup_{\Gamma} \inf_{\Delta} P(t_0, x_0, \Gamma, \Delta) \\ W^+(t_0, x_0) &= \inf_{\Delta} \sup_{\Gamma} P(t_0, x_0, \Gamma, \Delta). \end{aligned} \tag{4.1}$$

The number $W^+(t_0, x_0)$ is called the upper value of the game with initial point (t_0, x_0) and $W^-(t_0, x_0)$ is called the lower value. It is always true that

$$W^-(t_0, x_0) \leq W^+(t_0, x_0). \tag{4.2}$$

If $W^-(t_0, x_0) = W^+(t_0, x_0)$ we denote this common value by $W(t_0, x_0)$ and say that the game has a value equal to $W(t_0, x_0)$.

Let A and B be two sets of real numbers. We say that $A \geq B$ if for every a in A and every b in B the inequality $a \geq b$ holds.

A pair of strategies (Γ^*, Δ^*) will be called *optimal strategies* or will be said to constitute a *saddle point* of the game if

$$P(t_0, x_0, \Gamma, \Delta^*) \leq P(t_0, x_0, \Gamma^*, \Delta^*) \leq P(t_0, x_0, \Gamma^*, \Delta) \tag{4.3}$$

for all strategies Γ for Player I and all strategies Δ for Player II. It follows from (4.1) and (4.2) that if a saddle point exists, then the game has a value and

$$W(t_0, x_0) = P(t_0, x_0, \Gamma^*, \Delta^*)$$

Analogous definitions of upper value, lower value, value and saddle point point hold for the games G_c and G_r.

5. Games of Fixed Duration

In this section we summarize the principal results presented in [2], [4], [5] and [6] as they pertain to games of fixed duration. Although we refer the reader to the above cited papers for proofs and details we will give an account of the order in which various results are obtained. The presentation here will incorporate some improvements over the results in the above cited papers.

We consider W^+ and W^- as functions of initial point (t,x) in $[T_1,T) \times R^n$. If we set $W^+(T,x) = g(x)$ and set $W^-(T,x) = g(x)$ and if Assumption I holds, then it can be shown that W^+ and W^- are uniformly continuous on bounded subsets of $[T_0,T] \times R^n$ and uniformly Lipschitz continuous in x on bounded subsets of $[T_0,T] \times R^n$.

We now return to the game with fixed initial point (t_0,x_0). Let

$$v^0 = W^+(t_0,x_0) \qquad v_0 = W^-(t_0,x_0).$$

Let $W^+(T,x) = g(x)$ and let $W^-(T,x) = g(x)$. Let

$$C(v^0) = \{(\tau,\xi) : t_0 \le \tau \le T,\ \xi \in R^n,\ W^+(\tau,\xi) \ge v^0\}$$
$$C(v_0) = \{(\tau,\xi) : t_0 \le \tau \le T,\ \xi \in R^n,\ W^-(\tau,\xi) \le v_0\}.$$

Note that since W^+ and W^- are continuous functions, the sets $C(v^0)$ and $C(v_0)$ are closed.

The next lemma is central in our development. It plays the role of the principle of optimality.

Lemma 5.1. Let (τ,ξ) be a point of $C(v_0)$ such that $\tau < T$. Let t_1 satisfy $\tau < t_1 < T$ and let u be any control for Player I on $[\tau,t_1]$. Then there exists a relaxed control $\zeta = \zeta(u)$ such that the relaxed trajectory $\psi(\ ,\tau,\xi,u,\zeta)$ has the property that $(t_1,\psi(t_1)) \in C(v_0)$.

Our next result, which is an easy consequence of Lemma 5.1, involves the lower Dini derivates of the functions W^+ and W^-. Let L be a real valued function defined on a set $[T_0, T) \times \mathcal{G}$, where $\mathcal{G}$ is an open set in R^n and let h be a vector in R^n. The upper Dini derivate of L at (t_0, x_0) in the direction h is denoted by $D^+L(t, x; 1, h)$ and is defined by

$$D^+L(t,x;1,h) = \limsup_{\delta \downarrow 0} \; [L(t+\delta,\; x+\delta h) - L(t,x)]\,\delta^{-1}.$$

The lower Dini derivate of L in the direction $(1, h)$ at the point (t, x) is denoted by $D^-L(t, x;\; 1, h)$ and is defined by

$$D^-L(t,x;1,h) = \liminf_{\delta \downarrow 0} [L(t+\delta,\; x+\delta h) - L(t,x)]\delta^{-1}.$$

If L is differentiable at (t, x) then for every h in R^n, $D^+L(t, x; 1, h) = D^-L(t, x; 1, h)$. If we denote this common value by $DL(t, x; 1, h)$ we have

$$DL(t,x;1,h) = L_t(t,x) + \langle L_x(t,x), h\rangle.$$

Define

$$F_z(t,x,y) = co\{f(t,x,y,z) : z \in Z\}$$

$$F_y(t,x,z) = co\{f(t,x,y,z) : y \in Y\},$$

where $co(A)$ denotes the convex hull of A.

Lemma 5.2. Let Assumption I hold. Then for each (t, x) in $[T_0, T) \times R^n$

$$\max_{y \in Y} \left[\min_{h \in F_z(t,x,y)} D^-W^-(t,x;1,h) \right] \le 0$$

$$\min_{z \in Z} \left[\max_{h \in F_y(t,x,z)} D^+W^+(t,x;1,h) \right] \ge 0.$$

If we set

$$H^+(t,x,p) = \min_{z \in Z} \max_{y \in Y} \langle p, f(t,x,y,z)\rangle$$

$$H^-(t,x,p) = \max_{y \in Y} \min_{z \in Z} \langle p, f(t,x,y,z)\rangle,$$

then Lemma 5.2 implies the following:

Lemma 5.3. The upper value W^+ is a viscosity subsolution of the Cauchy problem

$$u_t + H^+(t, x, \nabla_x u) = 0$$

for (t, x) in $[T_0, T) \times R^n$ and

$$u(T, x) = g(x).$$

The lower value W^- is a viscosity supersolution of the Cauchy problem

$$u_t + H^-(t, x, \nabla_x u) = 0$$
$$u(T, x) = g(x).$$

For the definitions of viscosity solution, subsolution and supersolution see [8].

We now introduce an important condition in the theory of differential games.

Definition 5.1. If for all (t, x) in $[T_0, T_1]$ and all p in R^n, $H^+(t, x, p) = H^-(t, x, p)$, we say that the *Isaacs condition* holds. We denote the common value by $H(t, x, p)$.

At this point we could, if we so desired, appeal to a comparison theorem of Crandall and Lions [9] and show that if the Isaacs condition holds, then $W^+ = W^-$ and the game has value. Since we also wish to establish the existence of saddle points in games of fixed duration when the Isaacs condition holds, and since that argument will also give the existence of value we will not appeal to the theory of viscosity solutions to establish the existence of value.

We now state the principal result concerning games of fixed duration. For the proofs see [2], [5], [7], [15].

Theorem 5.1. Let Assumption I hold and let the Isaacs condition hold. Then the game of fixed duration has value $W(t, x)$ for each initial position (t, x) in $[T_0, T) \times R^n$. The value function W is Lipschitz continuous on bounded subsets of $[T_0, T) \times R^n$. If $W(T, x) \equiv g(x)$, then W is Lipschitz continuous on bounded subsets of $[T_0, T] \times R^n$ and W is the unique viscosity solution of the Isaacs equation

$$\begin{aligned} &u_t + H(t, x, \nabla_x u) = 0 \qquad (t, x) \in [T_0, T) \times R^n \\ &u(T, x) = g(x). \end{aligned} \tag{5.1}$$

Moreover, for each initial point (t_0, x_0) the game has a saddle point (Γ^*, Δ^*), which may depend on (t_0, x_0).

Corollary 5.1. At points of differentiability, the function W satisfies

$$W_t(t,x) + H(t,x,\nabla_x W(t,x)) = 0.$$

An examination of the arguments used to establish Theorem 5.1 show that if we consider the game G_c, in which constant valued controls are chosen over each interval, then Theorem 5.1 holds for the game G_c. If we denote the value of this game by W_c, then it follows from a uniqueness Theorem of Ishi **[12]** for viscosity solutions of (5.1) that $W_c = W$. This observation is useful when we approximate the game by its n-th stage discrete approximation.

If the Isaacs condition does not hold, then we use relaxed strategies and consider the game G_R. Let $\mathcal{Y}$ denote the set of probability measures on Y and let $\mathcal{Z}$ denote the set of probability measures on Z. Let

$$H_R^+(t,x,p) = \min_{dz \in \mathcal{Z}} \max_{dy \in \mathcal{Y}} \int_Z \int_Y \langle p, f(t,x,y,z)\rangle dy\, dz$$
$$H_R^-(t,x,p) = \max_{dy \in \mathcal{Y}} \min_{dz \in \mathcal{Z}} \int_Y \int_Z \langle p, f(t,x,y,z)\rangle dz\, dy$$

Then the Isaacs condition holds in the sense that

$$H_R^+(t,x,p) = H_R^-(t,x,p).$$

We denote this common value by $H_R(t,x,p)$. It follows that the analogue of Theorem 5.1 holds for the game G_R, with value function W_R.

Remark 5.1. If the game as originally defined satisfies the Isaacs condition

$$H^+(t,x,p) = H^-(t,x,p),$$

then

$$H(t,x,p) = H_R(t,x,p).$$

Hence W and W_R are viscosity solutions of (5.1). It then follows from the uniqueness theorem of Ishi [12], that $W = W_R$.

For a detailed discussion of the game G_R, see [4].

Other definitions of a differential game have been proposed by Friedman, Fleming and Elliott and Kalton. See [10] for these definitions. It has been shown by various authors that for games of fixed duration all of these values are viscosity solutions of (5.1). It then again follows from Ishi's uniqueness theorem [12] that all of these values are equal and are equal to our value.

6. Games of Generalized Pursuit and Evasion

In this section we summarize our results for games of generalized pursuit and evasion. For details and proofs see [3], [4].

We make the following assumption concerning the terminal set $\mathcal{T}$.

Assumption II. Let $\mathcal{T}$ be as in (2.2) and let $\mathcal{T}_1 = \mathcal{T} \cap \{[T_0, T_1) \times R^n\}$. Let $\partial\mathcal{T}_1$, the boundary of $\mathcal{T}_1$, be a $C^{(2)}$ manifold. At each point (t, x) of $\partial\mathcal{T}_1$ let (ν_0, ν) denote the normal to $\partial\mathcal{T}_1$ that points to the exterior of $\mathcal{T}_1$. Then

$$\min_z \max_y [\nu_0 + \langle \nu, f(t, x, y, z)\rangle] < 0, \tag{6.1}$$

where the min is taken over Z and the max is taken over Y.

For games of generalized pursuit and evasion the Isaacs condition is stated in terms of

$$\mathcal{H}(t, x, p^0, p, y, z) = p^0 f^0(t, x, y, z) + \langle p, f(t, x, y, z)\rangle.$$

Definition 6.1. The Isaacs condition is said to hold for a game of generalized pursuit and evasion if for all (t, x) in $[T_0, T_1]$ that are not in $\mathcal{T}$,

$$\max_y \min_z \mathcal{H}(t, x, 1, p, y, z) = \min_z \max_y \mathcal{H}(t, x, 1, p, y, z),$$

where the max is taken over Y and the min is taken over Z. In this event the common value is denoted by $\overline{H}(t,x,p)$.

We now state the principal result concerning games of generalized pursuit and evasion. Let

$$\mathcal{R} = \{[T_0, T_1] \times R^n\} - \mathcal{T}.$$

Theorem 6.1. Let Assumptions I and II hold and let the Isaacs condition of Definition (6.1) hold. For all (t,x,y,z) in $[T_0,T_1] \times R^n \times Y \times Z$ let $f^0(t,x,y,z) \geq \psi(t)$, where ψ is a function that is integrable on $[T_0,T_1]$. Then for each (t,x) in $\mathcal{R}$ the game with initial point (t,x) has value $W(t,x)$. The function W is Lipschitz continuous on compact subsets of $\mathcal{R}$ and is a viscosity solution of the Cauchy problem

$$u_t + \overline{H}(t,x,\nabla_x u) = 0 \qquad \text{on } \mathcal{R}$$

$$u(t,x) = 0 \qquad \text{on } \partial\mathcal{T}.$$

We note that we cannot assert that saddle points exist. If the Isaacs condition fails to hold, then we consider relaxed controls and obtain a theorem analogous to Theorem 6.1 for relaxed controls.

In conclusion we point out that the condition $f^0 \geq \psi$ can be replaced by the condition

$$\min_{y \in Y} \max_{z \in Z} \left[\nu_0 + \langle \nu, f(t,x,y,z)\rangle\right] < 0. \tag{6.2}$$

at all points of $\partial\mathcal{T}_1$.

7. Games of Survival

The principal result for games of survival is given in the following theorem.

THEOREM 7.1. *Let Assumptions I and II hold and let (6.2) hold. Let the Isaacs condition as stated in Definition (6.1) hold. Then for each (t,x) in $\mathcal{R}$ the game with initial* *point (t,x) has value $W(t,x)$. The function W is Lipschitz continuous on compact subsets of $\mathfrak{R}$ and is a viscosity solution of the Cauchy problem*

$$u_t + \overline{H}(t,x,\nabla_x u) = 0 \quad \text{on} \quad \mathfrak{R}$$
$$u(t,x) = g(x) \qquad\qquad \text{on} \quad \partial\mathfrak{T}.$$

The comments concerning the existence of saddle point and relaxed controls made after the statement of Theorem 6.1 also apply here. For details see [3], [4], [5], [15].

REFERENCES

[1] L. D. BERKOVITZ, *Lectures on differential games*, in "Differential Games and Related Topics," North–Holland, Amsterdam, London, 1971, pp. 3–45. H. W. Kuhn and G. P. Szegö, eds.

[2] L. D. BERKOVITZ, *The existence of value and saddle point in games of fixed duration*, SIAM J. Control **23** (1985), 172–196.

[3] L. D. BERKOVITZ, *Differential games of generalized pursuit and evasion*, SIAM J. Control **24** (1986), 361–373.

[4] L. D. BERKOVITZ, *Differential games without the Isaacs condition*, in "Recent Advances in Communication and Control Theory," Optimization Software, New York, 1987, pp. 308–336. R. E. Kalman, G. I. Marchuk, A. E. Roberts and A. J. Viterbi, eds.

[5] L. D. BERKOVITZ, *Characterizations of the values of differential games*, Appl. Math. Optim. **17** (1988), 177–183.

[6] L. D. BERKOVITZ, *Differential games of survival*, J. Math. Anal. Appl. **129** (1988), 493–504.

[7] L. D. BERKOVITZ, *Correction to The existence of value and saddle point in games of fixed duration*, SIAM J. Control **26** (1988), 740–742.

[8] M. G. CRANDALL, L. C. EVANS AND P. L. LIONS, *Some properties of viscosity solutions of Hamilton–Jacobi equations*, Trans. Amer. Math. Soc. **282** (1984), 487-502.

[9] M. G. Crandall and P. L. Lions, *Remarks on the existence and uniqueness of unbounded viscosity solutions of Hamilton–Jacobi equations*, Illinois J. Math. **31** (1987), 665–687.

[10] A. Friedman, *Differential games*, CBMS Regional Conference Series in Mathematics No. 18, American Mathematical Society, Providence, (1974).

[11] R. Isaacs, "Differential Games," John Wiley and Sons, New York, London, Sydney, 1965.

[12] H. Ishi, *Uniqueness of unbounded viscosity solutions of Hamilton–Jacobi equations*, Indiana Univ. Math. J. **33** (1984), 721–748.

[13] N. N. Krasovskii and A. I. Subbotin, "Positional Differential Games," Nauka, Moscow, 1974. (In Russian.)

[14] J. Warga, "Optimal Control of Differential and Functional Equations," Academic Press, New York, 1972.

[15] J. Yong, *On the Isaacs equation of differential games of fixed duration*, J. Optim. Theory Appl. **50** (1986), 359–364.

Contingent Isaacs Equations of a Differential Game

JEAN-PIERRE AUBIN

CEREMADE, UNIVERSITÉ DE PARIS-DAUPHINE

Abstract

The purpose of this paper is to characterize classical and lower semicontinuous solutions to the Hamilton-Jacobi-Isaacs partial differential equations associated with a differential game and, in particular, characterize closed subsets the indicators of which are solutions to these equations. For doing so, we replace the classical concept of derivative by contingent epiderivative, which can apply to any function. The use of indicator of subsets which are solutions of either one of the contingent Isaac allows to characterize areas of the playability set in which some behavior (playability, winability, etc.) of the players can be achieved.

Résumé

Le but de cet article est de caractériser les solutions classiques aussi bien que semi-continues inférieurement des diverses équations d'Hamilton-Jacobi-Isaacs associées à un jeu différentiel et, en particulier, de caractériser les ensembles fermés dont la fonction indicatrice est solution de l'une de ces équations. On doit pour cela remplacer le concept de dérivée par celui d'epidérivée contingente, qui peut s'appliquer à toute fonction. L'utilisation des fonctions indicatrices qui sont des solutions de l'une de ces équations contingentes d'Isaacs permet d'identifier les sous-ensembles caractéristiques de certains comportements des joueurs.

1 Introduction

Let us consider a differential game described by

$$\begin{cases} i) & z'(t) = h(z(t),u(t),v(t)) \\ ii) & u(t) \in U(z(t)) \\ iii) & v(t) \in V(z(t)) \end{cases} \tag{1}$$

where $h : \mathbf{R}^n \times \mathbf{R}^p \times \mathbf{R}^q \to \mathbf{R}^n$ describes the dynamics of the game and where the set-valued $U : \mathbf{R}^n \rightsquigarrow \mathbf{R}^p$ and $V : \mathbf{R}^n \rightsquigarrow \mathbf{R}^q$ are the a priori feedback maps of our players Xavier and Yves which represent the **state-dependent constraints** bearing on the controls of each player.

We denote by K, the playability set, the intersection of the domains of U and V.

The purpose of this paper is to characterize the solutions Φ to the following four Hamilton-Jacobi-Isaacs partial differential equations associated with this differential game.

$$\begin{cases} i) & \inf_{u\in U(z)} \inf_{v\in V(z)} \frac{d\Phi(z)}{dz} \cdot h(z,u,v) = 0 \\ ii) & \sup_{u\in U(z)} \sup_{v\in V(z)} \frac{d\Phi(z)}{dz} \cdot h(z,u,v) = 0 \\ iii) & \sup_{v\in V(z)} \inf_{u\in U(z)} \frac{d\Phi(z)}{dz} \cdot h(z,u,v) = 0 \\ iv) & \inf_{u\in U(z)} \sup_{v\in V(z)} \frac{d\Phi(z)}{dz} \cdot h(z,u,v) = 0 \end{cases}$$

We shall study the properties of the solutions (classical of lower semicontinuous) to these partial differential equations, and in particular, characterize the solutions which are indicators of closed subsets L, defined by

$$\Psi_L(x) := \begin{cases} 0 & \text{if } z \in L \\ +\infty & \text{if } z \notin L \end{cases}$$

which are only lower semicontinuous. For that purpose, we shall weaken the concept of usual derivative and replace it by the concept of contingent epiderivative, which can apply to any function[1], and generalize these Isaacs partial differential equation to corresponding contingent Isaacs equations[2].

The use of indicator of subsets which are solutions of either one of the contingent Isaac's equation allows to characterize areas of the playability set in which some behavior (playability, winability, etc.) of the players can be achieved.

Example Let us consider our two players, Xavier and Yves. Xavier acts on a state space X and Yves on a state space Y. For doing so, they have access to some knowledge about the global state (x,y) of the system and are allowed to choose controls u in a global state dependent set $U(x,y)$ and v in a global state dependent set $V(x,y)$ respectively.

[1]See [2, Chapter VII] for an introduction to nonsmooth and set-valued analysis

[2]In the extent where Isaacs are partial differential equations, they have, under adequate assumptions, unique "viscosity solutions", which are only continuous (See [3,4,8,9,10,20,21] and the references of theses papers). In the case of control problems, it has been shown in [12,13,14] that any viscosity solution is a solution to an adequate contingent version of Hamilton-Jacobi-Bellman equation. The comparison of the solutions in the case of differential games remains to be done.

Their actions on the state of the system are governed by the system of differential inclusions:

$$\begin{cases} a) & \begin{cases} i) & x'(t) = f(x(t),y(t),u(t)) \\ ii) & u(t) \in U(x(t),y(t)) \end{cases} \\ b) & \begin{cases} i) & y'(t) = g(x(t),y(t),v(t)) \\ ii) & v(t) \in V(x(t),y(t)) \end{cases} \end{cases} \tag{2}$$

We now describe the influences (power relations) that Xavier exerts on Yves and vice-versa through **rules of the game**. They are set-valued maps $P : Y \leadsto X$ and $Q : X \leadsto Y$ which are interpreted in the following way. When the state of Yves is y, Xavier's choice is constrained to belong to $P(y)$. In a symmetric way, the set-valued map Q assigns to each state x the set $Q(x)$ of states y that Yves can implement[3].

Hence, the **playability domain** of the game is the subset $K \subset X \times Y$ defined by:

$$K := \{ (x,y) \in X \times Y \mid x \in P(y) \text{ and } y \in Q(x) \} \tag{3}$$

Naturally, we must begin by providing sufficient conditions implying that the playability domain is non empty. Since the playability domain is the subset of fixed-points (x,y) of the set-valued map $(x,y) \leadsto P(y) \times Q(x)$, we can use one of the many fixed point theorems to answer this type of questions[4].

From now on, we shall assume that the playability domain associated with the rules P and Q is not empty.

By denoting by $z := (x,y) \in Z := X \times Y$ the global state, by $h(z,u,v) := (f(x,u,v), g(y,u,v))$ the values of the map $h : \mathbf{R}^n \times \mathbf{R}^p \times \mathbf{R}^q \to \mathbf{R}^n$ describing the dynamics of the game, by $L := \text{Graph}(P)$ Xavier's closed domain, by $M := \text{Graph}(Q^{-1})$ Yves's one and by $K := L \cap M$ the playability domain. We shall also identify the set-valued maps U and V with their restrictions to L and M respectively by setting $U(z) := \emptyset$ whenever $z \notin L$ and $V(z) := \emptyset$ when $z \notin M$. □

2 Contingent Isaacs Equations

Since we want to include indicators of subsets among the solutions of Isaacs equations and also, look for smaller lower semicontinuous solutions to such an equation satisfying such or such property, we are led to weaken the concept of usual derivatives involved in these partial differential equations by replacing them by contingent epiderivatives, since any extended function $\Phi : X \to \mathbf{R} \cup \{+\infty\}$ has contingent epiderivative[5], and in particular, indicators, for which we have the relation

$$D_\uparrow \Psi_L(z)(v) = \Psi_{T_L(z)}(v) := \begin{cases} 0 & \text{if } v \in T_L(z) \\ +\infty & \text{if } v \notin T_L(z) \end{cases}$$

[3] We can easily extend the results below to the time-dependent case using the methods of [1].

[4] For instance, Kakutani's Fixed Point Theorem furnishes such conditions: *Let $L \subset X$ and $M \subset Y$ be compact convex subsets and $P : M \leadsto L$ and $Q : L \leadsto M$ be closed maps with nonempty convex images. Then the playability domain is not empty.*

[5] the contingent epiderivative of Φ at $z \in \text{Dom}(\Phi)$ in the direction v is defined by

$$D_\uparrow \Phi(z)(v) := \liminf_{h \to 0+, u \to v} (\Phi(x + hu) - \Phi(x))/h$$

It is characterised by the fact that its epigraph is the contingent cone to the epigraph of Φ at $(z, \Phi(z))$.

Theorem 2.1 *Let us assume at least that $h : \mathbf{R}^n \times \mathbf{R}^p \times \mathbf{R}^q \to \mathbf{R}^n$ is continuous, has linear growth , and that the set-valued maps are closed with linear growth.*

We assume that the all extended functions Φ are nonnegative and contingently epidifferenti and that their domains are contained in the intersection K of the domains of U and V.

1 — If the values of the set-valued maps U and V are convex and if h is affine with respect to the controls, Φ is a solution to the contingent equation

$$\inf_{u\in U(x)} \inf_{v\in V(x)} D_\uparrow\Phi(z)(h(z,u,v)) = 0 \tag{4}$$

if and only if

$$\forall z \in \Phi,\ \exists z(\cdot) \in S(z) \mid \forall t \geq 0,\ \Phi(z(t)) \leq \Phi(z)$$

2 — Assume that h is uniformly lipschitzean with respect to x. Then Φ is a solution to the contingent equation

$$\sup_{u\in U(x)} \sup_{v\in V(x)} D_\uparrow\Phi(z)(h(z,u,v)) = 0 \tag{5}$$

if and only if

$$\forall z \in \mathrm{Dom}(\Phi),\ \forall\ z(\cdot) \in S(z), \forall t \geq 0,\ \Phi(z(t)) \leq \Phi(z)$$

3 — Assume that V is lower semicontinuous, that the values of U and V are convex and that h is affine with respect to u. Then Φ is a solution to the contingent equation

$$\sup_{v\in V(x)} \inf_{u\in U(x)} D_\uparrow\Phi(z)(h(z,u,v)) = 0 \tag{6}$$

if and only if for any continuous closed-loop control $\tilde{v}(z) \in V(z)$ played by Yves and any initial state $z \in \mathrm{Dom}(\Phi)$, there exists a solution $z(\cdot)$ to Xavier's control problem

$$\begin{cases} i) & z'(t) = h(z(t), u(t), \tilde{v}(z(t))) \\ ii) & u(t) \in U(z(t)) \end{cases} \tag{7}$$

starting at z and satisfying $\forall t \geq 0,\ \Phi(z(t)) \leq \Phi(z)$.

4 — Assume that V is lower semicontinuous with convex values. Then Φ is a solution to the contingent equation

$$\inf_{u\in U(x)} \sup_{v\in V(x)} D_\uparrow\Phi(z)(h(z,u,v)) = 0 \tag{8}$$

if and only if Xavier can play a closed-loop control $\tilde{u}(z) \in U(z)$ such that, for any continuous closed-loop control $\tilde{v}(z) \in V(z)$ played by Yves and for any initial state $z \in \mathrm{Dom}(\Phi)$, there exists a solution $z(\cdot)$ to

$$z'(t) = h(z(t), \tilde{u}(z(t)), \tilde{v}(z(t))) \tag{9}$$

starting at z and satisfying $\forall t \geq 0,\ \Phi(z(t)) \leq \Phi(z)$. The converse is true if

$$B_\Phi := \{\bar{u} \in U(z) \ :\ \sup_{v\in V(x)} D_\uparrow\Phi(z)(h(z,\bar{u},v)) = \inf_{u\in U(x)} \sup_{v\in V(x)} D_\uparrow\Phi(z)(h(z,u,v))\}$$

is lower semicontinuous with closed convex values.

[6]This means that for all $z \in \mathrm{Dom}(\Phi)$, $\forall\ v \in X$, $D_\uparrow\Phi(z)(v) > -\infty$ and that $D_\uparrow\Phi(z)(v) < \infty$ for at least a $v \in X$.

Proof — It is based on the properties of lower semicontinuous Lyapunov functions and universal Lyapunov functions of a differential inclusion which are stated in the appendix.

— The two first statements are translations of the theorems characterizing Lyapunov and universal Lyapunov functions applied to the differential inclusion $z'(t) \in H(z(t))$ where $H(z) := f(z, U(z), V(z))$.

— Let us prove the third one. Assume that Φ satisfies the stated property. Since V is lower semicontinuous with convex values, Michael's Theorem implies that for all $z_0 \in \text{Dom}(V)$ and $v_0 \in V(z_0)$, there exists a continuous selection $\tilde{v}(\cdot)$ of V such that $v(z_0) = v_0$. Then Φ enjoys the Lyapunov property for the set-valued map $H_{\tilde{v}}(z) := h(z, U(z), \tilde{v}(z))$, and thus, there exists $u_0 \in U(z_0)$ such that

$$D_{\uparrow}\Phi(z_0)(h(z_0, u_0, \tilde{v}(z_0))) \leq 0$$

Hence Φ is a solution to (6).

Conversely, assume that Φ is a solution to (6). Then for all closed-loop control $\tilde{v}$, the set-valued map $H_{\tilde{v}}$ satisfies the assumptions of the theorem characterizing Lyapunov functions, so that there exists a solution to the inclusion $z' \in H_{\tilde{v}}(z)$ for all initial state $z \in \text{Dom}(\Phi)$ satisfying $\forall\, t \geq 0,\ \ \Phi(z(t)) \leq \Phi(z)$.

— Consider finally the fourth statement. Assume that Xavier's can find a continuous closed-loop control $\tilde{u}$ such that for all closed-loop control $\tilde{v}$, Φ enjoys the stated property. Since V is lower semicontinuous with convex values, Michael's Theorem implies that for all $z_0 \in \text{Dom}(V)$ and $v_0 \in V(z_0)$, there exists a continuous selection $\tilde{v}(\cdot)$ of V such that $v(z_0) = v_0$. Since for any continuous closed-loop control $\tilde{v}(\cdot)$, Φ enjoys the Lyapunov property for the single-valued map $z \to h(z, \tilde{u}(z), \tilde{v}(z))$, we deduce that for all $z_0 \in \text{Dom}(\Phi)$, that there exists $u := \tilde{u}(z)$ such that for all $v \in V(z)$, $D_{\uparrow}\Phi(z)(h(x, u, v)) \leq 0$, so that Φ is a solution to (6).

Conversely, assume that the set-valued map B_{Φ} is lower semicontinuous with closed convex values. Hence Michael's Theorem implies that there exists a continuous selection $\tilde{u}$ of B_{Φ}. Then for any continuous closed-loop control $\tilde{v}(\cdot) \in V(\cdot)$, we deduce from (8) that Φ is a Lyapunov function for the single-valued map $z \to h(z, \tilde{u}(z), \tilde{v}(z))$, so that, for all $z \in \text{Dom}(\Phi)$, there exists a solution $z(\cdot)$ to the system (9) satisfying $\forall\, t \geq 0,\ \ \Phi(z(t)) \leq \Phi(z)$. □

3 Characterization of some behavioral properties

Let L be a closed subset of the intersection K of the domains of U and V. The problem we investigate is to find that a (or all) solution(s) $z(\cdot)$ of the game which is (are) viable in L. There are several ways to achieve that purpose, according to the cooperative or non cooperative behavior of the players. We shall examine here six of them.

Definition 3.1 *We shall say the a subset L enjoys:*

1 — the "playability property" if and only if

$$\forall\, z \in L,\ \exists\, z(\cdot) \in S(z) \mid \forall\, t \geq 0,\ \ z(t) \in L$$

2 — the "winability property" if and only if

$$\forall\, z \in L,\ \forall\ z(\cdot) \in S(z), \forall\, t \geq 0,\ \ z(t) \in L$$

3 — "Xavier's discriminating property" if and only if for any continuous closed-loop control $\tilde{v}(z) \in V(z)$ played by Yves and any initial state $z \in L$, there exists a solution $z(\cdot)$ to Xavier's control problem

$$\begin{cases} i) & z'(t) = h(z(t), u(t), \tilde{v}(z(t))) \\ ii) & u(t) \in U(z(t)) \end{cases} \tag{10}$$

starting at z and which is viable in L.

4 — "Xavier's leading property" if and only if Xavier can play a closed-loop control $\tilde{u}(z) \in U(z)$ such that, for any continuous closed-loop control $\tilde{v}(z) \in V(z)$ played by Yves and for any initial state $z \in L$, there exists a solution $z(\cdot)$ to (9) starting at z and viable in L.

We shall characterize these properties: for that purpose we associate with L the following set-valued maps:

— The regulation map R_L defined by

$$\forall\, z \in L,\ R_L(z) := \{\ (u,v) \in U(z) \times V(z) \mid h(z,u,v) \in T_L(z)\ \}$$

— Xavier's discriminating map A_L defined by

$$\forall\, z \in L,\ A_L(z,v) := \{\ u \in U(z) \mid (u,v) \in R_L(z)\ \}$$

— Xavier's leading map B_L defined by

$$\forall\, z \in L,\ \ B(z)\ :=\ \bigcap_{v \in V(z)} A_L(z,v)$$

Definition 3.2 *We shall say that*

— *L is playability domain if $\forall\, z \in L,\ R_L(z)\ \neq\ \emptyset$*

— *L is a winability domain if*

$\forall\, z \in L,\ R_L(z) := U(z) \times V(z)$

— *L is a Xavier's discriminating domain if*

$$\forall\, z \in L, \forall\, v \in V(z),\ A_L(z,v)\ \neq\ \emptyset \tag{11}$$

— *L is a Xavier's leading domain if $\forall\, z \in L,\ B_L(z)\ \neq\ \emptyset$*

We begin by translating these properties in terms of contingent Isaacs equations:

Proposition 3.1 *Let us assume that $h : \mathbf{R}^n \times \mathbf{R}^p \times \mathbf{R}^q \to \mathbf{R}^n$ is continuous, has linear growth , and that the set-valued maps are closed with linear growth.*

— *L is playability domain if and only if Ψ_L is a solution to (4)*

— *L is a winability domain if and only if Ψ_L is a solution to (5)*

— *L is a discriminating domain for Xavier if and only if Ψ_L is a solution to (6)*

— *L is a leading domain for Xavier if and only if Ψ_L is a solution to (8)*

Therefore, Theorem 2.1 implies the following characterization of these domains:

Corollary 3.1 *Let us assume at least that $h : \mathbf{R}^n \times \mathbf{R}^p \times \mathbf{R}^q \to \mathbf{R}^n$ is continuous, has linear growth , and that the set-valued maps are closed with linear growth.*

1 — If the values of the set-valued maps U and V are convex and if h is affine with respect to the controls, then L enjoys the playability property if and only if it is a playability domain.

2 — Assume that h is uniformly lipschitzean with respect to x. Then L enjoys the winability property if and only if it is a winability domain.

3 — Assume that V is lower semicontinuous, that the values of U and V are convex and that h is affine with respect to u. Then L enjoys Xavier's discriminating property if and only if it is a discriminating domain for Xavier.

4 — Assume that V is lower semicontinuous with convex values. If L enjoys Xavier's leading property, then it is a leading domain for him. The converse is true if B_L is lower semicontinuous with closed convex values.

The existence theorems of the viability and invariance kernels imply the following consequence:

Proposition 3.2 *Let us assume that $h : \mathbf{R}^n \times \mathbf{R}^p \times \mathbf{R}^q \to \mathbf{R}^n$ is continuous, has linear growth , and that the set-valued maps are closed with linear growth.*

1 — If the values of the set-valued maps U and V are convex and if h is affine with respect to the controls, then there exists a largest closed playability domain contained in L, whose indicator is the smallest lower semicontinuous solution to (4) larger than or equal to the indicator Ψ_L of L.

2 — Assume that h is uniformly lipschitzean with respect to x. Then there exists a largest closed winability domain contained in L, whose indicator is the smallest lower semicontinuous solution to (5) larger than or equal to the indicator Ψ_L of L.

Remark — The question whether there are largest closed discriminating and leading domains remains an open question. □

4 Appendix: Lower Semicontinuous Lyapunov Functions

We consider now a differential inclusion

$$\text{for almost all } t \geq 0,\ x'(t) \ \in \ F(x(t)) \tag{12}$$

and time-dependent functions $w(\cdot)$ defined as solutions to a differential equation

$$w'(t) = -\phi(w(t)),\ w(0) \ = \ V(x(0)) \tag{13}$$

where $\phi : \mathbf{R}_+ \to \mathbf{R}$ is a given continuous function with linear growth. This function ϕ is used as a parameter in what follows.

The main instance of such a function ϕ is the affine function $\phi(w) := aw - b$, the solutions of which are $w(t) = (w(0) - \frac{b}{a})e^{-at} + \frac{b}{a}$.

Our problem is to characterize either ϕ -Lyapunov functions, i.e., nonnegative extended functions $V : X \to \mathbf{R}_+ \cup \{+\infty\}$ satisfying

$$\forall t \geq 0, \;\; V(x(t)) \;\leq\; w(t), \;\; w(0) = V(x(0)) \tag{14}$$

along at least a solution to the differential inclusion (12) or ϕ- universal Lyapunov functions, wich satisfy property (14) along all solutions to (12).

Definition 4.1 *We shall say that a nonnegative contingently epidifferentiable*[7] *extended function V is a* **Lyapunov function** *of F associated with a function $\phi(\cdot) : \mathbf{R}_+ \mapsto \mathbf{R}$ if and only if V is a solution to the contingent Hamilton-Jacobi inequalities*

$$\forall\, x \in \mathrm{Dom}(V), \;\; \inf_{v \in F(x)} D_\uparrow V(x)(v) + \phi(V(x)) \;\leq\; 0 \tag{15}$$

and a **universal Lyapunov function** *of F associated with a function ϕ if and only if V is a solution to the upper contingent Hamilton-Jacobi inequalities*

$$\forall\, x \in \mathrm{Dom}(V), \;\; \sup_{v \in F(x)} D_\uparrow V(x)(v) + \phi(V(x)) \;\leq\; 0 \tag{16}$$

Theorem 4.1 *Let V be an nonnegative contingently epidifferentiable extended function and $F : X \leadsto X$ be a nontrivial set-valued map.*

— *Let us assume that F is upper semicontinuous with compact convex images and linear growth. Then V is a Lyapunov function of F associated with $\phi(\cdot)$ if and only if for all initial state $x_0 \in Dom(V)$, there exist solutions $x(\cdot)$ to differential inclusion (12) and $w(\cdot)$ to differential equation (13) satisfying property (14).*

— *If F is lipschitzean on the interior of its domain with compact values, then V is a universal Lyapunov function associated with ϕ if and only if for all initial state $x_0 \in Dom(V)$, all solutions $x(\cdot)$ to differential inclusion (12) and $w(\cdot)$ to differential equation (13) do satisfy property (14).*

The proof is based on the viability and invariance theorems of the closed subset EpV for the differential inclusion:

$$\begin{cases} i) & x'(t) \in F(x(t)) \\ ii) & w'(t) = -\phi(w(t)) \end{cases} \tag{17}$$

and these viability and invariance theorems can be reformulated the in the following way:

Corollary 4.1 *Let $F : X \leadsto X$ be a nontrivial set-valued map.*

— *Let us assume that F is upper semicontinuous with compact convex images and linear growth.*

A closed subset K enjoys the viability property if and only if its indicator Ψ_K is a solution to the contingent equation

$$\inf_{v \in F(x)} D_\uparrow \Psi_K(x)(v) \;=\; 0$$

[7] This means that for all $x \in \mathrm{Dom}(V)$, $\forall\, v \in X$, $D_\uparrow V(x)(v) > -\infty$ and that $D_\uparrow V(x)(v) < \infty$ for at least a $v \in X$.

— *If F is lipschitzean on the interior of its domain with compact values, then K is invariant by F if and only if its indicator Ψ_K is a solution to the contingent equation*

$$\sup_{v\in F(x)} D_\uparrow\Psi_K(x)(v) = 0$$

The functions ϕ and $U : X \to \mathbf{R}_+ \cup \{+\infty\}$ being given, can we construct the smallest lower semicontinuous Lyapunov function of a set-valued map F associated to ϕ larger than or equal to U, i.e., the smallest nonnegative lower semicontinuous solution U_ϕ to the contingent Hamilton-Jacobi inequalities (15) larger than or equal to U?

Theorem 4.2 *Let us consider a nontrivial set-valued map $F : X \rightsquigarrow X$, a continuous function $\phi : \mathbf{R}_+ \to \mathbf{R}$ with linear growth and a proper nonnegative extended function U.*

— *Let us assume that F is upper semicontinuous with compact convex images and linear growth. Then there exists a smallest nonnegative lower semicontinuous solution $U_\phi : Dom(F) \mapsto \mathbf{R} \cup \{+\infty\}$ to the contingent Hamilton-Jacobi inequalities (15) larger than or equal to U (which can be the constant $+\infty$), which then enjoys the property:*

$$\begin{array}{ll}\forall\, x \in \mathrm{Dom}(U_\phi), & \text{there exists solutions to (13) and (14) satisfying} \\ & \forall\, t \geq 0,\ \ U(x(t)) \leq U_\phi(x(t)) \leq w(t)\end{array}$$

— *If F is lipschitzean on the interior of its domain with compact values and ϕ is lipschitzean, then there exists a smallest nonnegative lower semicontinuous solution $\tilde{U}_\phi$: $Dom(F) \mapsto \mathbf{R} \cup \{+\infty\}$ to the upper contingent Hamilton-Jacobi inequalities (13) larger than or equal to U (which can be the constant $+\infty$), which then enjoys the property:*

$$\begin{array}{ll}\forall\, x \in \mathrm{Dom}(U_\phi), & \text{all solutions to (13) and (14) satisfy} \\ & \forall\, t \geq 0,\ \ U(x(t)) \leq U_\phi(x(t)) \leq w(t)\end{array}$$

In particular, for $\phi(w) := aw$, we deduce that

$$\forall\, x \in \mathrm{Dom}(U_a),\ U(x(t)) \leq U_a(x_0)e^{-at} \ \text{ and thus, converges to } 0$$

The proof amounts to show that the largest closed viability domain (invariance domain) contained in the epigraph of U, called the viability kernel (invariance kernel) of Ep(U), which does exist under the assumptions of the first (second) part of the theorem, is actually an epigraph, and thus, the one of the smallest lower semicontinuous (universal) Lyapunov function. Actually, the existence theorems of these kernels are equivalent to the above theorem, since it implies the following

Corollary 4.2 *We posit the assumptions of Theorem 4.2.*

— *Let us assume that F is upper semicontinuous with compact convex images and linear growth.*

The indicator $\Psi_{\mathrm{Viab}(K)}$ of the viability kernel Viab(K) *of a closed subset K (i.e., the largest closed viability domain of F contained in K) is the smallest nonnegative lower semicontinuous solution to*

$$\forall\, x \in \mathrm{Dom}(V),\ \inf_{v\in F(x)} D_\uparrow V(x)(v) \leq 0 \tag{18}$$

larger than or equal to Ψ_K.

— *Assume that* F *is lipschitzean on the interior of its domain with compact values.*

The indicator $\Psi_{\mathrm{Inv}(K)}$ *of the invariant kernel* $\mathrm{Inv}(K)$ *of a closed subset* K *(i.e., the largest closed invariance domain of* F *contained in* K*) is the smallest nonnegative lower semicontinuous solution to*

$$\forall\, x \in \mathrm{Dom}(V), \quad \sup_{v \in F(x)} D_{\uparrow}V(x)(v) \;\leq\; 0 \tag{19}$$

larger than or equal to Ψ_K.

References

[1] AUBIN J.-P. & CELLINA A. (1984) DIFFERENTIAL INCLUSIONS. Springer-Verlag (Grundlehren der Math. Wissenschaften, Vol.264, 1-342)

[2] AUBIN J.-P. & EKELAND I. (1984) APPLIED NONLINEAR ANALYSIS. Wiley-Interscience

[3] BARDI M.(1988) *This volume*

[4] BERKOWITZ L. (1988) *This volume*

[5] BERNHARD P. (1979) Contribution à l'étude des jeux différentiels à somme nulle et information parfaite. Thèse Université de Paris VI

[6] BERNHARD P. (1980) *Exact controllability of perturbed continuous-time linear systems.* Trans. Automatic Control, 25, 89-96

[7] BERNHARD P. (1987) In Singh M. G. Ed. SYSTEMS & CONTROL ENCYCLOPEDIA, Pergamon Press

[8] CRANDALL M. G., & LIONS P.L., (1983) *Viscosity solutions of Hamilton-Jacobi equations* . Trans. Amer. Math. Soc., 277, 1-42

[9] CRANDALL M.G., , EVANS L.C., & LIONS P.L., (1984) *Some properties of viscosity solutions of Hamilton-Jacobi equation* . Trans. Amer. Math. Soc., 282(2), 487-502

[10] FLEMMING W. (1988)

[11] FLEMMING W. & RISHEL R.W. (1975) DETERMINISTIC AND SOCHASTIC OPTIMAL CONTROL Springer-Verlag

[12] FRANKOWSKA H. (1987) *L'équation d'Hamilton-Jacobi contingente.* Comptes Rendus de l'Académie des Sciences, PARIS,

[13] FRANKOWSKA H. *Optimal trajectories associated to a solution of contingent Hamilton-Jacobi Equation* Appl. Math. Opt.

[14] FRANKOWSKA H. (to appear) *Hamilton-Jacobi Equations:viscosity solutions and generalized gradients*. J. Math.Anal. Appli.

[15] GUSEINOV H. G. , SUBBOTIN A. I. & USHAKOV V. N. (1985) *Derivatives for multivalued mappings with applications to game theoretical problems of control*. Problems of Control and Information Theory, Vol.14, 155-167

[16] ISSACS R. (1965) DIFFERENTIAL GAMES. Wiley, New York

[17] KRASOVSKI N. N. (1986) THE CONTROL OF A DYNAMIC SYSTEM. Nauka, Moscow

[18] KRASOVSKI N. N. & SUBBOTIN A. I. (1974) POSITIONAL DIFFERENTIAL GAMES. Nauka, Moscow

[19] LEITMANN G. (1980) *Guaranteed avoidance strategies*. Journal of Optimization Theory and Applications, Vol.32, 569-576

[20] LIONS P. -L. (1982) GENERALIZED SOLUTIONS OF HAMILTON-JACOBI EQUATIONS. Pitman

[21] LIONS P. -L. & SOUGANIDIS P.E. (1985) *Differential games, optimal control and directional derivatives of viscosity solutions of Bellman and Isaacs' equations*. SIAM J. Control. Optimization, 23

[22] SUBBOTIN A. I. (1985) *Conditions for optimality of a guaranteed outcome in game problems of control*. Proceedings of the Steklov Institute of Mathematics, 167, 291-304

[23] SUBBOTIN A. I. & SUBBOTINA N. N. (1983) *Differentiability properties of the value function of a differential game with integral terminal costs*. Problems of Control and Information Theory, 12, 153-166

[24] SUBBOTIN A. I. & TARASYEV A. M. (1986) *Stability properties of the value function of a differential game and viscosity solutions of Hamilton-Jacobi equations*. Problems of Control and Information Theory, 15, 451-463

A PDE FRAMEWORK FOR GAMES OF PURSUIT-EVASION TYPE

Martino BARDI & Pierpaolo SORAVIA

Dipartimento di Matematica Pura e Applicata, Universita' di Padova

via Belzoni 7, I-35131 Padova, Italy

Abstract. The capture time T of a pursuit-evasion game is considered. Under a "capturability" condition it is shown that T is the unique viscosity solution of the Isaacs equation coupled with certain singular boundary conditions. Moreover the upper and lower values of the game are compared by PDE methods, as well as the corresponding "capturability" sets, and, under the Isaacs condition, it is shown that the game has a value.

1. Introduction.

We consider a dynamical system controlled by two players

$$\begin{cases} y' = f(y,a,b) & , y \in \mathbb{R}^N, a \in A, b \in B, \\ y(0) = x \end{cases} \tag{1.1}$$

where $A, B \subseteq \mathbb{R}^M$ are compact sets, f is continuous. We will study the game where the first player seeks to minimize the time taken by the system to reach a closed target set $\mathcal{T}$, while the second player wants to maximize it. We will associate with this problem lower and upper value functions T and $\tilde{T}$, each defined and finite in a suitable "capturability" set $\mathcal{R}$ and $\tilde{\mathcal{R}}$, respectively.

If we define the Hamiltonian

$$H(x,p) := \inf_{b\in B} \sup_{a\in A} \{ -f(x,a,b)\cdot p - 1 \},$$

then the boundary value problem

$$\begin{cases} H(x,DT) = 0 & , \text{ in } \mathcal{R} \setminus \mathcal{T} \\ T = 0 & , \text{ on } \partial\mathcal{T} \\ T(x) \to +\infty & , \text{ as } x \to \partial\mathcal{R} \end{cases} \tag{1.2}$$

is naturally associated, at least formally, to the capture time T.

Following Evans-Souganidis [11] and Evans-Ishii [9] , it is not hard to show that, when T is continuous, it satisfies the Hamilton-Jacobi-Isaacs equation in the viscosity sense (see Crandall-Evans-Lions [5] , Crandall-Lions [7] , Lions [17]). In Section 2 we introduce a "capturability" condition under which T is locally Lipschitz continuous and it takes up the boundary values in (1.2) uniformly.

The main result of this paper states that T is indeed the unique viscosity solution of (1.2). The analogous result for the classical minimum time problem in control theory has been recently proved by Bardi [1].

The first step in the uniqueness proof in [1] is the Kruzkov ' s change of variable [16]

$v(x) := \lambda^{-1}(1-e^{-\lambda T(x)})$, $\lambda>0$,

which transforms (1.2) into

$$\begin{cases} \lambda v + H(x,Dv) = 0 & , \text{ in } \Omega \setminus \mathcal{T} \\ v = 0 & , \text{ on } \partial\mathcal{T} \\ v = \lambda^{-1} & , \text{ on } \partial\Omega \end{cases} \tag{1.3}$$

with $\Omega = \mathcal{R}$. In Section 3 we give a uniqueness result for (1.3) for an arbitrary open set $\Omega \supset \mathcal{T}$, which implies the desired uniqueness for (1.2).

Now we observe that, if we extend $v(x) = \lambda^{-1}$ off $\mathcal{R}$, then (1.3) is satisfied with $\Omega = \mathbb{R}^N$. Therefore one can consider $\mathcal{R}$ as an unknown set, try to solve (1.3) with $\Omega = \mathbb{R}^N$, and then recover both T and $\mathcal{R}$ via the formulas

$T(x) = -\lambda^{-1}\log(\ 1-\lambda v(x)\ )$, $\mathcal{R} = \{\ x : v(x) < \lambda^{-1}\ \}$.

This point of view has been introduced for control problems by Bardi-Falcone [2], where the convergence of discrete approximations is studied. Here we apply this remark to prove that $\tilde{\mathcal{R}} \subset \mathcal{R}$ and $T \le \tilde{T}$, and to show that, when the Isaacs condition holds, $\tilde{\mathcal{R}} = \mathcal{R}$ and $T = \tilde{T}$, i. e. the game has a value.

For a general introduction to differential games of pursuit-evasion type and more references see Friedman [12], Hajek [13], Krassovski-Soubbotine [15], Rodin [19]. Some recent contributions to the theory are due to Berkovitz [3,4] , Lions-Souganidis [18]. The minimum time problem in control theory has been recently studied by viscosity solutions methods by Hermes [14] and Evans-James [10] as well.

2. A capturability condition and the regularity of T.

Let $f : \mathbb{R}^N \times A \times B \to \mathbb{R}^N$ and A, B, $\mathcal{T}$ have the properties mentioned in Section 1. We define the set of admissible controls

$\mathcal{A} := \{\ a \mid a : \mathbb{R}_+ \to A$ measurable $\}$, $\mathcal{B} := \{\ b \mid b : \mathbb{R}_+ \to B$ measurable $\}$,

the set of nonanticipating strategies for the first player

$\Delta := \{\ \alpha : \mathcal{B} \to \mathcal{A} \mid b(t)=\tilde{b}(t)\ \forall\ t \le \tilde{t}$ implies $\alpha[b](t) = \alpha[\tilde{b}](t)\ \forall\ t \le \tilde{t}\ \}$,

while the set Γ of nonanticipating strategies for the second player is defined in an obvious symmetric way. We denote by $y_x(s;a,b)$, $s \ge 0$, or simply by $y_x(s)$ or $y(s)$, the solution of (1.1) corresponding to the controls a, b and the initial state x, and by

$t_x = t_x(a,b) := \inf\ \{\ t : y(t) \in \mathcal{T}\ \} \le +\infty.$

We now introduce the lower and the upper values of the game following

Elliott-Kalton [8] :

$$T(x) := \inf_{\alpha\in\Delta} \sup_{b\in\mathcal{B}} t_x(\alpha[b],b) \ , \quad \tilde{T}(x) := \sup_{\beta\in\Gamma} \inf_{a\in\mathcal{A}} t_x(a,\beta[a]) \ .$$

Now set $\mathcal{R}(t) := \{ x \mid T(x)<t \}$, $\mathcal{R} :=\{ x \mid T(x)<+\infty \}$, $\tilde{\mathcal{R}} :=\{ x \mid \tilde{T}(x)<+\infty \}$, $d(x) := \mathrm{dist}(x,\mathcal{T})$, $X_\eta := \{ x : \mathrm{dist}(x,X)<\eta \}$. Our assumptions are

$$|f(x,a,b)-f(z,a,b)| \le L|x-z| \quad \forall x,z \in \mathbb{R}^N, a\in A, b\in B; \tag{H1}$$

$$\begin{cases} \exists\, \delta, \xi > 0 \text{ such that , setting } \mathcal{C} := \mathcal{T}_\delta \setminus \operatorname{int} \mathcal{T}, \\ \text{(i) } d \in C^2(\mathcal{C}), \\ \text{(ii) } \forall x\in\mathcal{C}\ \exists\, a_x \text{ such that } \quad f(x,a_x,b)\cdot Dd(x) \le -\xi \quad \forall b \in \mathcal{B}. \end{cases} \tag{H2}$$

To motivate the capturability condition (H2) we remark that if $\mathcal{T}$ is compact (ii) is equivalent to

$$f(x,a_x,b)\cdot n(x) < 0 \qquad \forall x \in \partial\mathcal{T} , b \in B, \tag{2.1}$$

where $n(x)$ is the exterior normal to $\mathcal{T}$. This last condition is well-known (see e. g. [13]) and we can prove that it is necessary for the Lipschitz continuity of T. For analogous controllability conditions in control theory see [2] and the references therein. We also note that (H1), the compactness of A and B, and the continuity of f, imply

$$|f(x,a,b)| \le K(1+|x|) \qquad \forall x \in \mathbb{R}^N. \tag{2.2}$$

<u>Proposition 1:</u> If (H1) ,(H2) hold, then

(i) $T(x) \le 2\xi^{-1}d(x)$ for all $x\in \mathcal{T}_\delta$,

(ii) $\mathcal{T} \subset \operatorname{int} \mathcal{R}(t)$ for all $t>0$,

(iii) $\mathcal{R}$ is open,

(iv) T is locally Lipschitz continuous in $\mathcal{R}$,

(v) for each $x_0 \in \partial\mathcal{R}$, $\lim_{x\to x_0} T(x) = +\infty$.

Proof. (i) For $x \in \mathcal{C}$ define $R(x) := (1+|x|)\exp(-2Kd(x)\xi^{-1})$,

$C_1 := \sup\{ |Dd(y)| : y \in B(x,R(x)) \cap \mathcal{C} \}$, $C_2 := \sup\{ |D^2 d(y)| : y \in B(x,R(x)) \cap \mathcal{C} \}$,

$C = (LKC_1(1+R(x)) + C_2K^2(1+R(x))^2)/2$,

and the nonanticipating strategy

$$\alpha[b](t) := \begin{cases} a(x) & \text{if } 0 \le t \le \xi/4C, \\ a(y(n\xi/4C)) & \text{if } n\xi/4C < t \le (n+1)\xi/4C,\ n=1,2,\ldots \end{cases}$$

(we denote here $a(x) := a_x$). It is easy to see, by an induction argument, that for all $t \le 2\xi^{-1}d(x) =: \tilde{t}$ we have $y(t) \in \mathcal{T}_\delta$ and

$$f(y(t),\alpha[b](t),b(t)) \cdot Dd(y(t)) \le -\xi/2 \quad \text{for all } b \in \mathcal{B},$$

which implies

$$d(y(\tilde{t})) \le d(x) - \tilde{t}\xi/2 = 0,$$

and completes the proof of (i).

(ii) is an immediate consequence of (i).

(iii) Fix $z \in \mathcal{R}$, $0 < \varepsilon \le 1$. There exists $\alpha \in \Delta$ such that for all $b \in \mathcal{B}$ we have $t_z(\alpha[b],b) \le T(z)+\varepsilon$. By (H1) and Gronwall's Lemma we have $|y(t_z)| \le (1+|z|)\exp(KT(z)+K) =: N$ and

$$|y_x(t) - y_z(t)| \le |x-z|\exp(Lt). \tag{2.3}$$

By (ii) we can take $0 < \sigma < \delta$ such that $(\mathcal{T} \cap B(0,N))_\sigma \subseteq \mathcal{R}$. Then $|x-z| \le \sigma\exp((-LT(z)-L)$ implies $y_x(t_z) \in \mathcal{R}$, which completes the proof of (iii).

(iv) We suppose $T(x) \ge T(z)$. The above argument and (i) show that

$$T(x) \le T(z) + \varepsilon + T(y_x(t_z)) \le T(z) + \varepsilon + 2\xi^{-1}\exp(LT(z)+L)|x-z| \tag{2.4}$$

and the local Lipschitz continuity follows at once.

(v) Fix $x \in \partial\mathcal{R}$ and assume by contradiction that there exists a sequence

$x_n \in \mathcal{R}$ such that $|x_n - x| \le 1/n$ and $T(x_n) \le M$ for all n. Choose $N=(2+|x_0|)\exp(KM+K)$ and σ as above. Fix $n \ge \exp(LM+L)/\sigma$. There exists a strategy α such that, for all $b \in \mathcal{B}$, $t_{x_n}(\alpha[b],b) \le M+1$. Therefore $y_x(t_{x_n}) \in (\mathcal{T} \cap B(0,N))_\sigma \subset \mathcal{R}$, which contradicts (iii). ///

Now we are going to prove a regularity result for the transform v we defined in Section 1, which we extend by setting $v(x)=\lambda^{-1}$ off $\mathcal{R}$. We observe that v is itself the lower value of a differential game with discount rate, i. e.

$$v(x) = \inf_{\alpha \in \Delta} \sup_{b \in \mathcal{B}} \int_0^{t_x} e^{-\lambda s} ds \ , \ \forall x \in \mathbb{R}^N, \tag{2.5}$$

and therefore it satisfies the following dynamic programming equation (see e. g. [9,11,18]) :

$$v(x) = \inf_{\alpha \in \Delta} \sup_{b \in \mathcal{B}} \left\{ \int_0^{\tau \wedge t_x} e^{-\lambda s} ds + e^{-\lambda \tau} v(y_x(\tau \wedge t_x)) \right\} \qquad \forall \tau > 0 .$$

<u>Proposition 2:</u> If (H1), (H2) hold and $\lambda > L$, then v is Lipschitz continuous in $\mathbb{R}^N$.

<u>Proof.</u> Fix $0 < \tau' < \tau$ and set $\varepsilon = \lambda^{-1}(1-e^{-\lambda \tau})$, $\varepsilon' = \lambda^{-1}(1-e^{-\lambda \tau'})$, $\Omega = \{ x \in \mathcal{R} : v(x) > \varepsilon \}$. Let ω, ω' be respectively the moduli of continuity of v in $\mathbb{R}^N$ and in Ω. By (2.4) T, and therefore v, are Lipschitz continuous in $\mathcal{R}(\tau)$, thus for C>0

$$\omega(r) \le \omega'(r) + Cr \tag{2.6}$$

Choose $x, z \in \Omega$, $v(x) \ge v(z)$ and let $0 < \eta \le \varepsilon - \varepsilon'$. By the dynamic programming principle there exists $\tilde{\alpha} \in \Delta$ such that for all $b \in \mathcal{B}$:

$$v(z) \ge \int_0^{\tau' \wedge t_z} e^{-\lambda s} ds + e^{-\lambda \tau'} v(y_z(\tau' \wedge t_z)) - \eta \tag{2.7}$$

and by (H2) we can assume that $y_z(t) \in \mathcal{T}$ for all $t>t_z$, for all $b \in \mathcal{B}$. Also there exists b_1 such that for $\tilde{\alpha}$ we have

$$v(x) \le \int_0^{\tau' \wedge t_x} e^{-\lambda s} ds + e^{-\lambda \tau'} v(y_x(\tau' \wedge t_x)) + \eta . \qquad (2.8)$$

It is easy to see that, by the choice of η, we have $t_x(\tilde{\alpha}[b_1],b_1) \ge \tau'$. If $t_z(\tilde{\alpha}[b_1],b_1) \ge \tau'$, (2.7), (2.8) give

$$v(x)-v(z) \le e^{-\lambda\tau'}(v(y_x(\tau'))-v(y_z(\tau'))) + 2\eta.$$

Otherwise we can take b_2 such that $t_z(\tilde{\alpha}[b_2],b_2) \ge \tau'$, so that

$$v(x)-v(z) \le e^{-\lambda\tau'}\big(v(y_x(\tau';\tilde{\alpha}[b_1],b_1)) - v(y_z(\tau';\tilde{\alpha}[b_2],b_2)) \big) + 2\eta \le$$

$$\le e^{-\lambda\tau'}\big(v(y_x(\tau';\tilde{\alpha}[b_1],b_1)) - v(y_z(\tau';\tilde{\alpha}[b_1],b_1)) \big) + 2\eta$$

because $v(y_z(\tau';\tilde{\alpha}[b_1],b_1)) = 0$. In any case we get $\omega'(r) \le e^{-\lambda\tau'}\omega(re^{L\tau'})+2\eta$, which implies, using (2.6)

$$\omega(r) \le e^{-\lambda\sigma}\omega(re^{L\sigma}) + Cr \qquad \text{for some } \sigma>0 . \qquad (2.9)$$

Given $0<r<1$, we fix $n \in \mathbb{N}$ such that $e^{-(n+1)L\sigma} \le r < e^{-nL\sigma}$ Then we compute, using $n+1$ times (2.9)

$$\omega(r) \le \omega(e^{-nL\sigma}) \le e^{-(n+1)\lambda\sigma}\omega(e^{L\sigma}) + C \sum_{j=1}^{n} e^{-(jL+(n-j)\lambda)\sigma} \le$$

$$\le C' e^{-(n+1)L\sigma} \le C'r ,$$

which concludes the proof. ///

3. Comparison and uniqueness results.

First, we give one more definition. We say that the function v takes the value γ (resp. $+\infty$) uniformly on a set X if for every $\varepsilon>0$ (resp. $M>0$) there exists $\eta>0$ such that $|v(x)-\gamma| \le \varepsilon$ (resp. $v(x) \ge M$) for all $x \in X_\eta$. All the results of this section are deduced from the following comparison theorem:

Theorem 1: Assume (H1). Let v_1 and v_2 be respectively an upper semicontinuous subsolution and a lower semicontinuous supersolution in the viscosity sense of (1.3)

$$\begin{cases} \lambda v + H(x,Dv) = 0 & , \text{ in } \Omega \setminus \mathcal{T}, \\ v = 0 & , \text{ on } \partial\mathcal{T}, \\ v = \lambda^{-1} & , \text{ on } \partial\Omega, \end{cases}$$

where Ω is an open neighbourhood of $\mathcal{T}$ and each v_i is bounded and takes up the boundary values uniformly. Then $v_1 \le v_2$. (It is understood that the second boundary condition disappears if $\Omega = \mathbb{R}^N$).

Proof. It is essentially the same as the proof of Theorem 2 in [1]. In fact that proof works for any Hamiltonian satisfying the estimate

$$|H(x,p)-H(z,q)| \le L|x-z|\,|p| + K(1+|z|)|p-q| ,$$

which is easily seen to hold in the present case. The possibility of comparing semicontinuous functions is remarked in [6]. ///

Corollary: Assume (H1), (H2). Then T is the unique viscosity solution of

$$\begin{cases} H(x,DT) = 0 & , \text{ in } \mathcal{R} \setminus \mathcal{T}, \\ T = 0 & , \text{ on } \partial\mathcal{T}, \\ T = +\infty & , \text{ on } \partial\mathcal{R}, \end{cases}$$

which is bounded below and takes up the boundary values uniformly.

Proof. From Propositions 1 and 2 it follows that T is continuous, it satisfies the boundary conditions uniformly, and $\mathcal{R}$ is open. The dynamic programming principle for T implies that T is a viscosity solution by standard arguments (e. g. [1,11,17]). The uniqueness is a consequence of Theorem 1 applied to the Kruzkov transform v, see [1, Thm. 2]. ///

Remark: Analogs to Propositions 1-2 and the Corollary hold for the upper value $\tilde{T}$, replacing $\mathcal{R}$ by $\tilde{\mathcal{R}}$ and H by $\tilde{H}$ defined as

$$\tilde{H} := \sup_{a\in A} \inf_{b\in B} \{ -f(x,a,b)\cdot p - 1 \} .$$

Theorem 2: Assume (H1), (H2). Then $\tilde{\mathcal{R}} \subset \mathcal{R}$ and $T(x) \le \tilde{T}(x)$ for all $x \in \tilde{\mathcal{R}}$. If moreover $H = \tilde{H}$ (Isaacs condition), then $\mathcal{R} = \tilde{\mathcal{R}}$ and $T = \tilde{T}$.

Proof. Since v is the lower value of a differential game (see (2.5)), it is a viscosity solution of (1.3) with $\Omega = \mathbb{R}^N$. On the other hand $\tilde{v} := \lambda^{-1}(1-e^{-\lambda \tilde{T}})$, extended to be $\tilde{v} = \lambda^{-1}$ off $\tilde{\mathcal{R}}$, is the upper value of the same game. Thus $\tilde{v}$ is a viscosity solution of (1.3) with $\Omega = \mathbb{R}^N$ and H replaced by $\tilde{H}$. Since $H \ge \tilde{H}$, $\tilde{v}$ is also a supersolution of (1.3) and all the conclusions follow from Theorem 1, using Proposition 2 and the Remark.

///

References.

[1] M. Bardi: A boundary value problem for the minimum time function, Siam J. Control , 1988 to appear.

[2] M. Bardi, M. Falcone: An approximation scheme for the minimum time function, preprint Universita' di Roma 1988.

[3] L.D. Berkovitz: Differential games of generalized pursuit and evasion, SIAM J. Control 24 (1986), 361-373.

[4] L.D. Berkovitz: Characterization of the values of differential games, Appl. Math. Optim. 17 (1988), 177-183.

[5] M.C. Crandall, L.C. Evans, P.L. Lions: Some properties of viscosity solutions of Hamilton-Jacobi equations, Trans. Amer. Math. Soc. 282 (1984), 487-502.

[6] M.C. Crandall, H. Ishii, P.L. Lions: Uniqueness of viscosity solutions revisited, J. Math. Soc. Japan 39 (1987), 581-596.

[7] M.C. Crandall, P.L. Lions: Viscosity solutions of Hamilton-Jacobi equations, Trans. Amer. Math. Soc. 277 (1984), 1-42.

[8] R.J. Elliott, N.J. Kalton: The existence of value in differential games, Mem. Amer. Math. Soc. 126 (1972).

[9] L.C. Evans, H. Ishii: Differential games and nonlinear first order PDE on bounded domains, Manuscripta Math. 49 (1984), 109-139.

[10] L.C. Evans, M.R. James: The Hamilton-Jacobi-Bellman equation for time-optimal control, preprint.

[11] L.C. Evans, P.E. Souganidis: Differential games and representation formulas for solutions of Hamilton-Jacobi equations, Indiana Univ. Math. J. 33 (1984), 773-797.

[12] A. Friedman: Differential Games, John Wiley, New York 1971.

[13] O. Hajek: Pursuit Games, Academic Press, New York 1975.

[14] H. Hermes: Feedback synthesis and positive, local solutions to Hamilton-Jacobi-Bellman equations, Proc. 1987 MTNS Conference, Sacks-Byrns-Martin Eds., North Holland, to appear.

[15] N. Krassovski, A. Soubbotine: Jeux Différentiels, Mir, Moscow 1977.

[16] S.N. Kruzkov: Generalized solutions of the Hamilton-Jacobi equations of the eikonal type I, Math. USSR Sbornik 27 (1975), 406-445.

[17] P.L. Lions: Generalized solutions of Hamilton-Jacobi equations, Pitman, Boston 1982.

[18] P.L. Lions, P.E. Souganidis: Differential games, optimal control and directional derivatives of viscosity solutions of Bellman' s and Isaacs' equations, SIAM J. Control 23 (1985), 566-583.

[19] E.Y. Rodin: A pursuit-evasion bibliography-version I, Comput. Math. Applic. 13 (1987), 275-340.

TIME-OPTIMAL PURSUIT INSIDE A CIRCLE

Professor John V. Breakwell
Dept. of Aeronautics/Astronautics, Stanford University
Durand Building - Rm. 220, Stanford, CA 94305, U.S.A.

ABSTRACT

Lion, having a speed advantage, chases Man inside a circular cage, the game terminating when their positions coincide. The solution to this game, computed for a speed-ratio 1.2, includes a "focal surface" of "radial" positions leading to curved motion by both players until termination, with Lion having to respond immediately to any change by Man between clockwise and anticlockwise motion. The solution also includes an "indifference zone" for Man from inside of which he must arrive at a particular point on the opposite side of the cage center from Lion by the time Lion reaches the center, curved motion by both then ensuing.

1. INTRODUCTION

Lion (L) and Man (M) move with constant speeds inside a circular cage, Lion the pursuer having a speed advantage. The game ends with Lion's position coincides with Man's. Clearly, from some "elementary" initial positions, Man runs directly away from Lion but is caught without ever reaching the perimeter of the cage. From other positions, as illustrated in Figure 1, Man runs toward the further (A) of the two points of intersection of the perimeter with the instantaneous Apollonius circle, i.e., the locus of position P such that $PL = w \cdot PM$, $w(> 1)$ being the speed-ratio. [The elementary positions correspond to an Apollonius circle entirely inside the cage.]

Now, as Lion and Man run towards A, the Apollonius circle shrinks toward zero, the directions *AM, AL* and *ML* remaining fixed. Equivalently, by continually enlarging the scale, we may regard *A, C, L, M* as fixed while the distance AO increases toward infinity. For any point A on the Apollonius circle the possible direction of the cage center O is restricted (see dashed lines in Figure 1) one limiting direction being due to the requirement that $OA \leq OM, OL$, and the other, parallel to ML, being due to the requirement that A *remain* the further of the two intersections of the perimeter with the Apollonius circle.

In particular, the Apollonius strategies cannot apply if O, L, M are collinear (with L between M and O) and with the Apollonius circle intersecting the perimeter in two distinct points. For such "radial" positions it is plausible that Lion will remain radial as he nears Man and that Man will apply at least part of his velocity perpendicular to the radius, being eventually caught on the perimeter. Note that Man can switch at any time between clockwise and anticlockwise motion, Lion having to respond immediately.

If we further assume that from positions almost but not quite radial Lion will move to a radial position, these radial positions will constitute a "focal surface" in the essentially 3-dimensional state space, approachable from either side. Otpimality requires that the approaching paths arrive tangentially.

The focal surface paths, which will turn out to be curved by both players, together with all their straight line tangential arrival paths, will be computed for a particular speed-ratio w and shown, together with the elementary and Apollonius positions, to fill up the appropriate state space.

The optimality of the completed solution will be demonstrated by appealing to the "viscosity condition."

2. ANALYSIS

The state variables (r, s, θ) and the controls (ϕ, ψ) are defined as in Figure 2. The main equation, τ denoting time-to-capture, is:

$$\underset{\phi \quad \psi}{\mathrm{MinMax}}\left\{1 + \tau_s \cos\psi + \tau_r w \cos\phi + \tau_\theta\left[\frac{\sin\psi}{s} - \frac{w \sin\phi}{r}\right]\right\} = 0 \tag{1}$$

so that

$$\frac{\cos\psi}{\tau_s} = \frac{\sin\psi}{(\tau_\theta/s)} = \frac{1}{\sqrt{\tau_s^2 + (\tau_\theta/s)^2}} \tag{2a}$$

$$\frac{\cos\phi}{-\tau_r} = \frac{\sin\phi}{(\tau_\theta/r)} = \frac{1}{\sqrt{\tau_r^2 + (\tau_\theta/r)^2}} \tag{2b}$$

and

$$H = 1 + \sqrt{\tau_s^2 + (\tau_\theta/s)^2} - w\sqrt{\tau_r^2 + (\tau_\theta/r)^2} = 0 \tag{3}$$

Along paths in the focal surface $\theta = 0$ we must have $\dot{\theta} = 0$, so that

$$\frac{1}{s^2\sqrt{\tau_s^2 + (\tau_\theta/s)^2}} = \frac{w}{r^2\sqrt{\tau_r^2 + (\tau_\theta/r)^2}} \tag{4}$$

and hence, because of (3)

$$\sqrt{\tau_r^2 + (\tau_\theta / r)^2} = \frac{ws^2}{w^2s^2 - r^2} \tag{5a}$$

$$\sqrt{\tau_s^2 + (\tau_\theta / s)^2} = \frac{r^2}{w^2s^2 - r^2} \tag{5a}$$

The discontinuity in the gradient of τ across the focal surface is clearly just a reversal of the sign of τ_θ. Along the focal surface paths, then:

$$\frac{d}{dt}(\tau_s) = -\partial H / \partial s$$

and hence

$$\frac{d}{dt}(\cos\psi) = \frac{1}{s} + \frac{(w^2s^2 + r^2)\cos^2\psi}{w^2s^2 - r^2} - \frac{2w^3s^2\cos\psi\cos\phi}{r(w^2s^2 - r^2)} \tag{6}$$

and, of course

$$\phi = \mathrm{Sin}^{-1}\left[\frac{r}{ws}\sin\psi\right] \tag{7}$$

3. NUMERICAL RESULTS

Units of distance and time have been chosen so that the cage radius as well as Man's speed are unity. Backwards integration of (6), together with (7) and $\dot{r} = w\cos\phi$, $\dot{s} = \cos\psi$, from capture:

$$r_f = 1 \ , \quad s_f = 1$$

for various ψ_f between 0 and $\pi/2$, determines the focal surface paths. These are shown in Figure 3 for $w = 1.2$. The paths terminate (in backward time) either at $s = 1$ with $\cos\psi < 0$ or else at $\cos\psi < \pm 1$ where τ_θ vanishes so that the gradient of τ becomes continuous. Figure 3 also shows the radial motion tributaries arriving tangentially at the terminations with $\cos\psi = \pm 1$. Note that if Man starts on the perimeter $(s = 1)$ with Lion in a radial position with $r > .3$ (approximately), Man moves inside the perimeter before being captured on the perimeter. If $0 < r < .3$ (approximately), Man starts by running radially inward towards Lion.

It is interesting to note that ϕ and ψ are continuous at termination with $\cos\psi = \pm 1$ although $\dot{\phi}$ and $\dot{\psi}$ are infinite. A similar situation arose in the backwards construction of a focal surface in a different problem [1], point-capture of two evaders in succession.

Here a single exceptional focal surface path reaches $r = 0$ in backward time with $s = s_0$, $\psi = \pi/2$, in fact with $\cos\psi \simeq r/(3ws_0)$ for small r, where s_0 depends on w. For $w = 1.2$, $s_0 \cong .835$, but this cannot be obtained directly from the backward integration which is unstable near $r = 0$; instead, forward integration must be used starting with small positive r, $s = s_0$ and $\cos\psi = r/(3ws_0)$, s_0 being adjusted so that r and s reach unity simultaneously. Note that τ_s and τ_θ both vanish at the point $r = 0$, $s = s_0$ and Man's arrival direction there is non-unique. As (barely) indicated by the τ–contours included in Figure 3, the value $s = s_0$ is the optimal radial distance of Man when Lion is at the center $(r = 0)$.

Figure 4 shows the curved paths of both players after Lion passes through the center, Man having chosen counterclockwise motion which, as also illustrated, he can reverse at any time.

The computation of the time-to-go function $\tau(r, s, \theta)$ is straightforward using straight line motions by both players arriving tangentially at various positions on the focal surface. The τ–contours for sections $r = 1$, .75, .5, and .1 of the state space are shown in Figures 5-8. For each r there is a circular "indifference zone," I with center at $s = s_0$, $\theta = \pi$, this zone shrinking to zero as $r \to 0$: from inside the indifference zone Man must reach $s = s_0$, $\theta = \pi$ by the time Lion reaches the center. The tangent motions arriving at $r = 1$, $s = 1$ coincide with the Apollonius motions of Figure 1 when OA is parallel to ML so that the final $\dot{\theta}$ is zero. This provides the outer boundaries of the Apollonius regions Ⓐ The circular boundaries between the elementary regions Ⓔ and the Apollonius regions Ⓐ correspond, of course, to A at position A_0 in Figure 1. For $r = .1$, region Ⓐ is too thin to be visible (in Figure 8).

4. PROOF OF OPTIMALITY

The form of the τ–contours in the neighborhood of the focal surface is sketched in Figure 9a. Figure 9b shows, in addition, a typical contour for a smooth approximation τ_s to τ, with $(\tau - \tau_s)$ always having a minimum on the focal surface $\theta = 0$. A sufficient condition, in addition to the main equation (3), for the optimality of the constructed function $\tau(r, s, \theta)$ is the "viscosity condition" [2]:

$$H(x, (\tau_s)_x) \leq 0 \qquad \text{at all minima of} \quad (\tau - \tau_s)$$

and all smooth approximations $\tau_s(r, s, \theta)$, x denoting the state (r, s, θ). We may suppose that τ_s coincides with τ along $\theta = 0$, in which case

$$(\tau_s)_r = \tau_r \quad , \quad (\tau_s)_s = \tau_s \qquad \text{but} \qquad |(\tau_s)_\theta| \le (\tau_\theta)$$

The viscosity condition thus becomes:

$$H_s(\alpha^2) \;=\; 1 + \sqrt{\tau_s^2 + \alpha^2(\tau_\theta/s)^2} - w\sqrt{\tau_r^2 + \alpha^2(\tau_\theta/r)^2} \ge 0 \tag{8}$$

for $0 \le \alpha^2 \le 1$. The validity of (8) is easily checked since $H_s(1) = 0$, $H'_s(1) = 0$ (equivalent to the tangency condition $\dot{\theta} = 0$) and $H'_s(\alpha^2) > 0$ for $\alpha^2 < 1$ (using the fact that $\sin\phi < \sin\psi$).

It is interesting to note that, after arrival at the focal surface $\theta = 0$, Lion faces a "perpetuated dilemma" until capture in having to respond immediately to any switch in direction by Man. Nevertheless, due to the tangency of the paths approaching the focal surface, we may show that if Lion responds after a small time-delay ε, the resulting increase in capture time $\to 0$ as $\varepsilon \to 0$. Strictly speaking we should also modify our definition of capture using coincidence to within some small distance of order ε. Figure 10 shows a typical path of Lion in the neighborhood of the radius OM. The non-tangential departures from this radius, after a switch by Man, have duration $O(\varepsilon)$, while the tangential returns have duration $O(\sqrt{\varepsilon})$. Hence, even though the number of possible switches can grow infinitely large as $\varepsilon \to 0$, the *proportion* of time spent by Lion in non-optimal play $\to 0$ as $\varepsilon \to 0$, and hence the time itself spent in non-optimal play $\to 0$. This proves our assertion.

5. THE INCLUSION OF RANDOM VELOCITY ERRORS

If we suppose that the velocity vectors of Lion and Man undergo small "white" random errors with the same noise-level, and if we again slightly modify our definition of capture, the time-to-capture τ^* satisfies:

$$\frac{q}{2}\left[\tau^*_{rr} + \tau^*_{ss} + \left(\frac{1}{r^2} + \frac{1}{s^2}\right)\tau^*_{\theta\theta}\right] + H(x, \tau^*_x) \;=\; 0 \quad , \tag{9}$$

where q indicates the noise-level. The resulting smooth function $\tau^*(r, s, \theta)$ satisfies, along $\theta = 0$:

$$\tau^*_\theta \;=\; 0 \quad , \quad \tau^*_{\theta\theta} > 0 \tag{10}$$

For sufficiently small q, $\tau^*_{\theta\theta}$ will be large enough to dominate the effect of τ^*_{rr}, τ^*_{ss} in (9). Thus $1 + d\tau^*/dt$ would be negative except for the presence of the diffusion term in (9).

However, if the *true* noise-level is less than that used in computing from (9) a solution τ^* (and the numerical computation becomes harder as q becomes smaller) then the corresponding computed strategy (ϕ^*, ψ^*) is non-optimal in favor of Lion: Man should not use $\psi = 0$ or π along the entire focal surface.

REFERENCES

[1] J.V. Breakwell and P. Hagedorn, "Point Capture of Two Evaders in Succession," *J.O.T.A.*, Vol. 27, No. 1, 1979.

[2] P. Bernhard, "Differential Games: Isaacs' Equation," *Encyclopedia of Systems and Control,* Editor: Madan Singh, Pergamon Press, 1987.

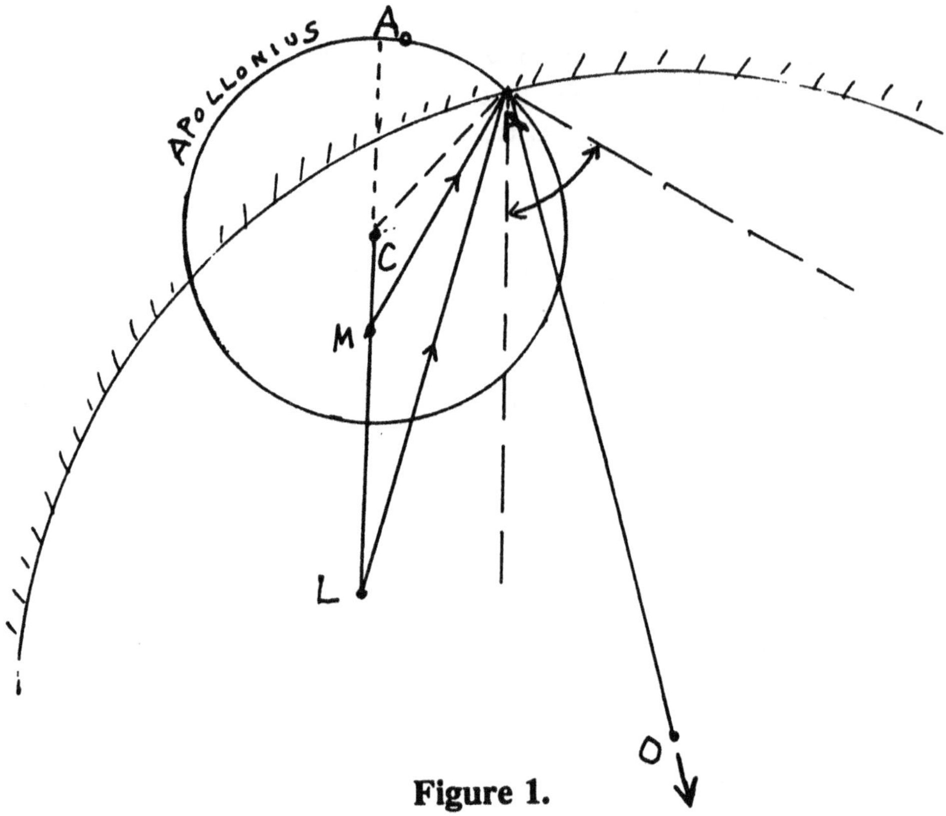

Figure 1.

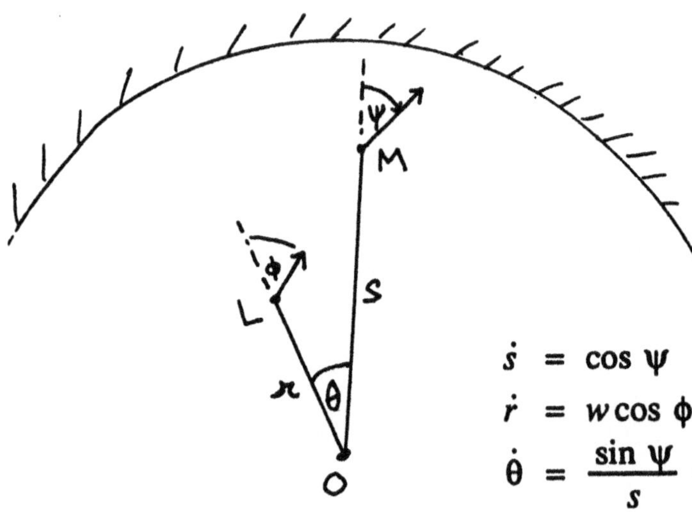

$$\dot{s} = \cos\psi$$
$$\dot{r} = w\cos\phi$$
$$\dot{\theta} = \frac{\sin\psi}{s} - \frac{w\sin\phi}{r}$$

Figure 2.

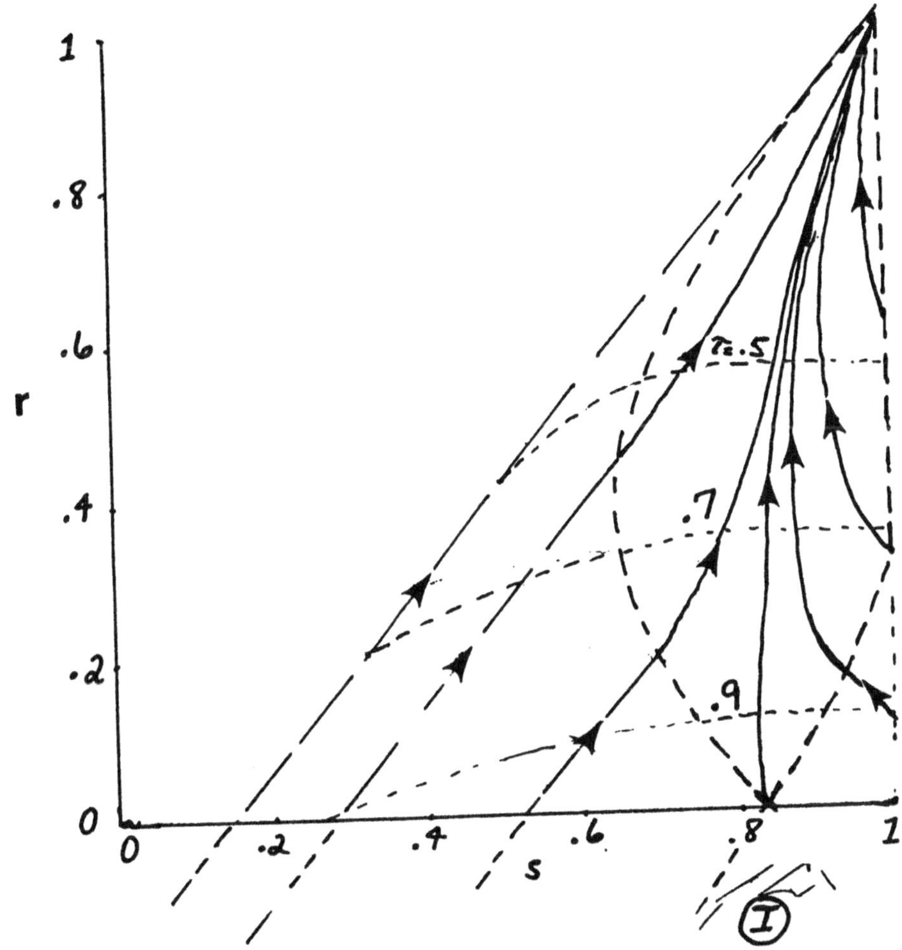

Focal Surface and Radial Tributaries
$w = 1.2$

Figure 3.

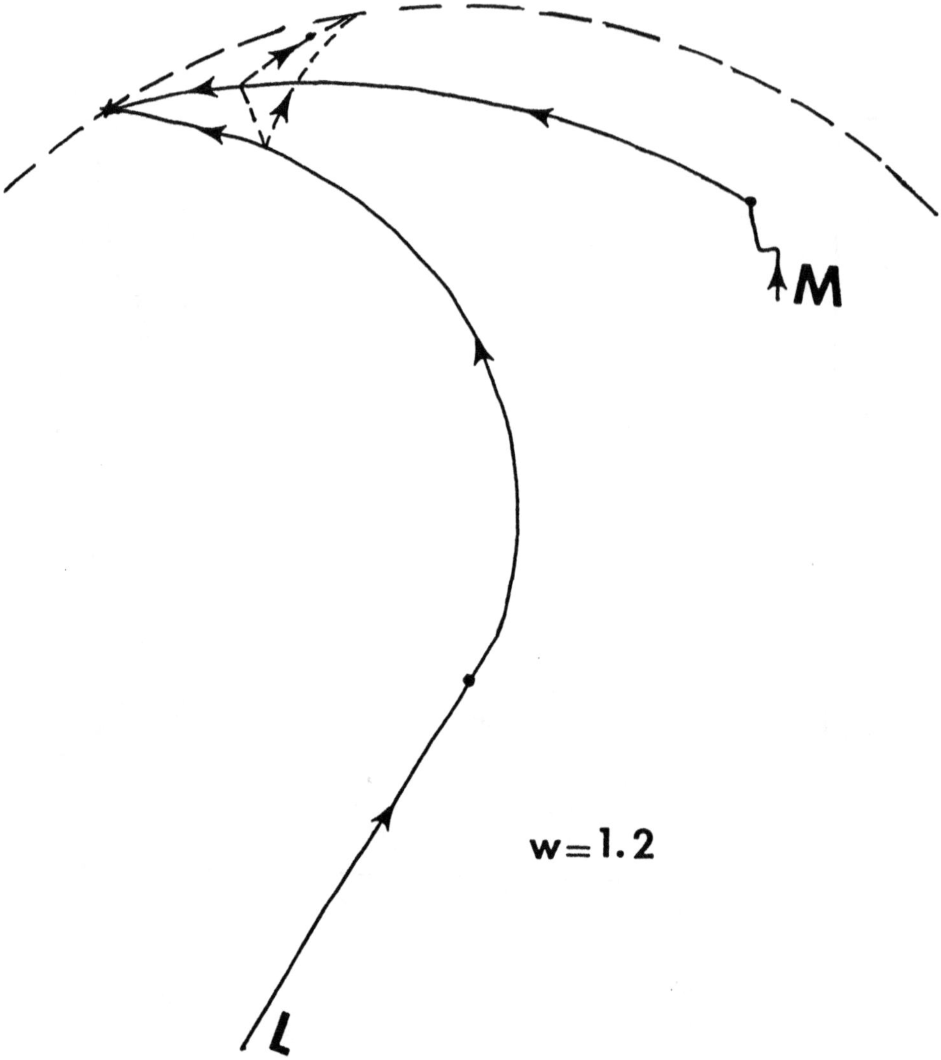

Path if Lion Passes Through Center

Figure 4.

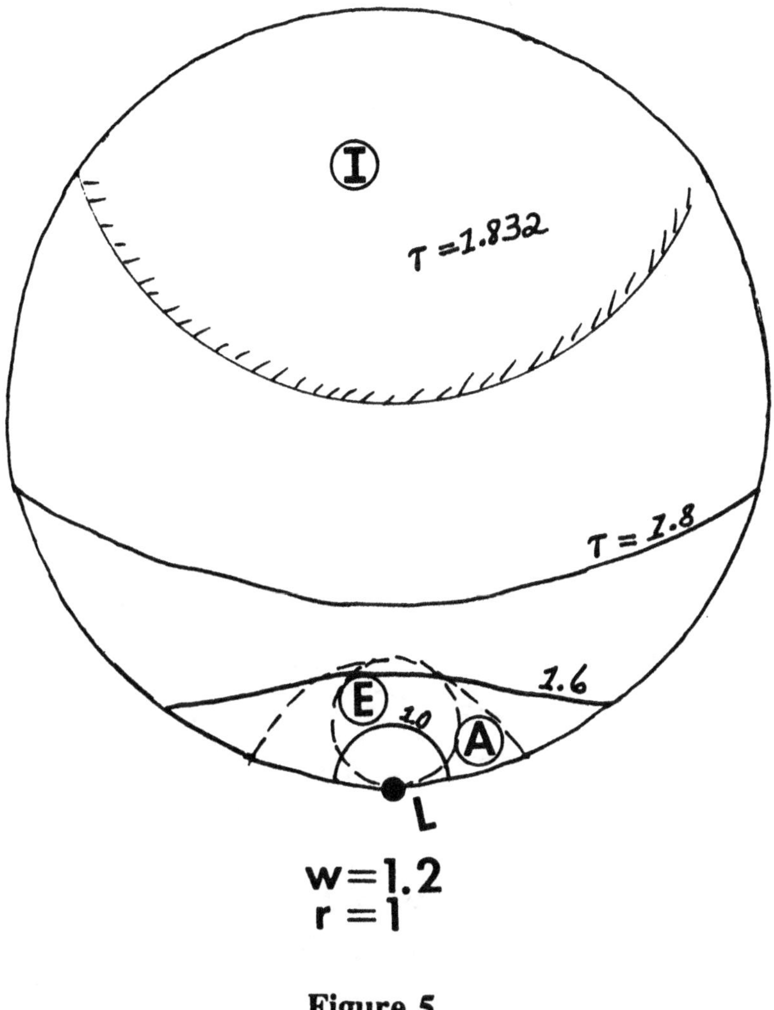

Figure 5.

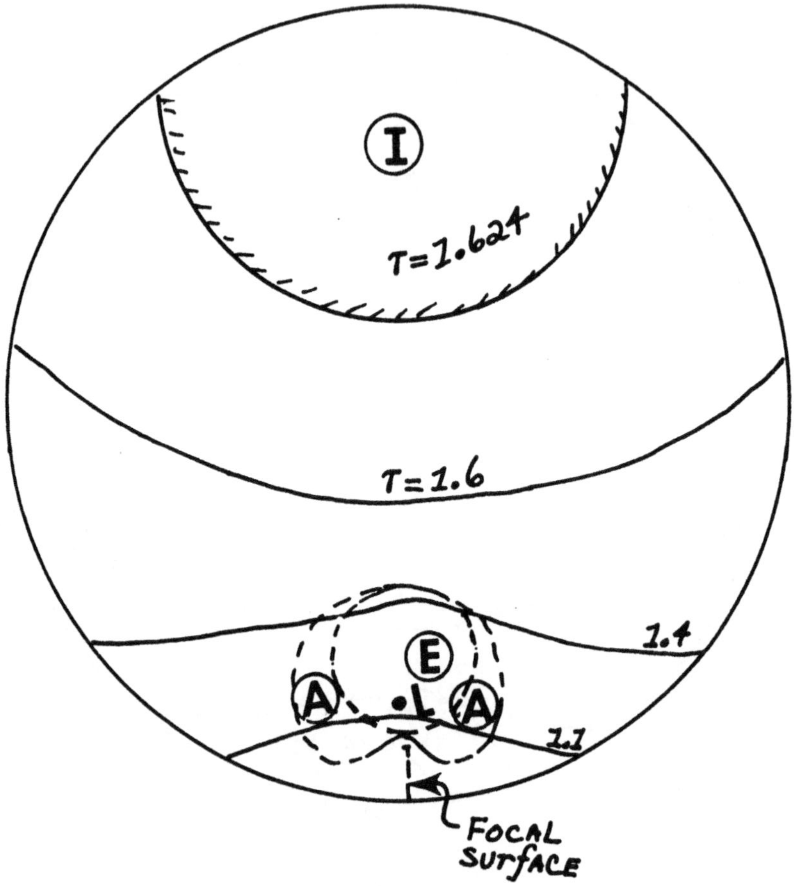

w = 1.2
r = .75

Figure 6.

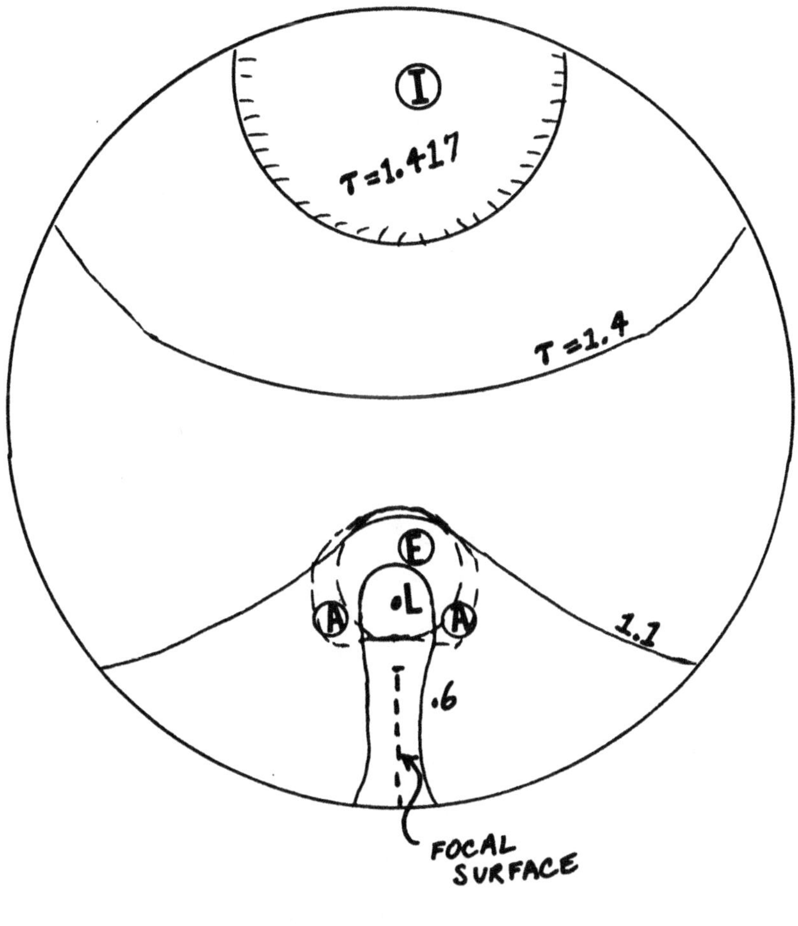

w = 1.2
r = .5

Figure 7.

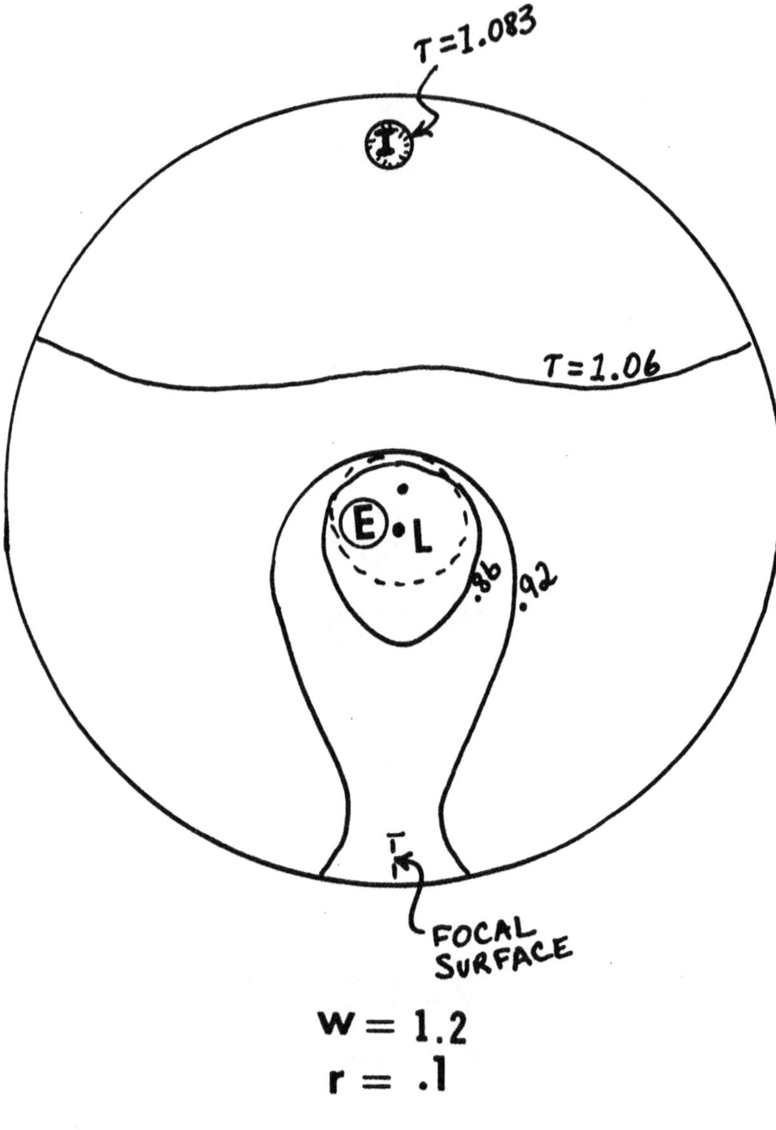

w = 1.2
r = .1

Figure 8.

Figure 9a.

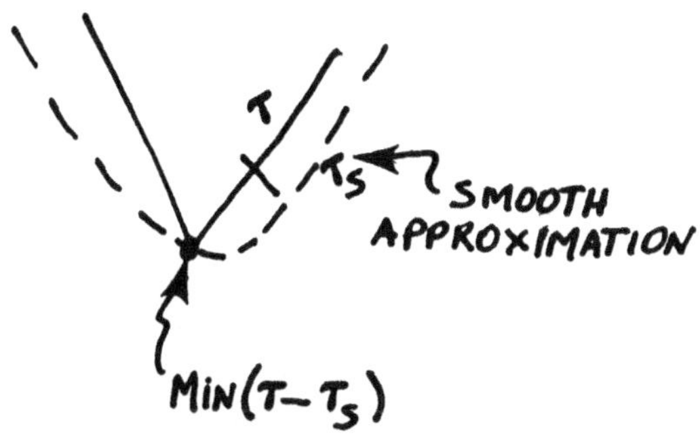

Figure 9b.

Figure 10.

A FEEDBACK GUIDANCE FOR PRE-LAUNCH MANEUVERING IN MEDIUM RANGE AIR COMBAT WITH MISSILES

U.R. Prasad[1], W. Grimm[2], E. Berger[3]
DFVLR Institute for Flight Systems Dynamics
Oberpfaffenhofen, 8031 Wessling, FRG

Dedicated to Henry J. Kelley†

Abstract

A new approach is presented for aircraft guidance in the pre-launch phase of medium range missile combat. Control actions are defined such that they work in a response mode to take advantage of the opponent's strategy errors. The method is based on miss distances resulting from hypothetical missile exchanges. Thus it is computationally expensive, since each evaluation of the guidance law requires a set of missile/target simulations. The performance of the proposed scheme is assessed in simulation encounters between two F-4 aircraft.

1. Introduction

The problem of medium range combat between two aircraft with one missile each has attracted some attention in the recent literature. Herbst (Ref.1) resorts to a piloted simulation approach to the problem. Both Järmark (Ref.2) and Moritz, Polis and Well (Ref.3) formulate the problem as a full order combat differential game and obtain open loop solutions to the quantitative game (or the game of degree). The payoff functions are generally chosen as weighted combinations of the miss distances in the two post-launch missile-aircraft encounters. However, the logic for the selection of weights by the players has not been developed completely. Meanwhile, Shinar and Davidovitz (Ref.4) address the qualitative game (or game of kind) with constant speed (reduced order) models. None of the above approaches deals with the determination of a feedback strategy to a combat player that is

---

[1]Visiting Scientist, on leave from Indian Institute of Science, Bangalore 560012, India

[2]Research Scientist, DFVLR

[3]Senior Programmer, DFVLR

either optimal or exploits the possible nonoptimal play by the other. In this paper a new approach for the problem of feedback guidance during the pre-launch maneuvering phase of medium range missile combat is presented. It is related to the ideas of threat reciprocity (Ref.5), reprisal strategies (Ref.6) and bicriterion games (Ref.7) and attempts to exploit the nonoptimal play by the opponent.

2. Problem Formulation

We consider two aggressive aircraft (A1 and A2) with one missile each (M1 and M2). Each vehicle satisfies a set of flight equations of the form

$$\dot{x} = V \cos\gamma \cos\chi \quad (1)$$

$$\dot{y} = V \cos\gamma \sin\chi \quad (2)$$

$$\dot{h} = V \sin\gamma \quad (3)$$

$$\dot{V} = [\xi T - D(h,V,n)]/m - g \sin\gamma \quad (4)$$

$$\dot{\gamma} = g\,(n \cos\mu - \cos\gamma)/V \quad (5)$$

$$\dot{\chi} = g\, n \sin\mu \,/\, (V \cos\gamma) \quad (6)$$

where (x,y,h) is the position vector (in Cartesian coordinates) and (V,γ,χ) the velocity vector (in polar coordinates) of the vehicle. The load factor n, the bank angle μ and the power setting ξ (set to 1 for missiles) are its control variables. The mass m is assumed constant for the aircraft and is a given function of time for the missiles. g is the gravitational acceleration. Maximum thrust T is a function of h and V for the aircraft; it is a given function of time for the missiles (decaying to 0 in the coasting phase).
The control constraints are given by

$$0 \le \xi \le 1 \quad (7)$$

$$n \le n_{max}\,(h,V) \quad (8)$$

where n_{max} is due to stall or structural limitations. The terrain and dynamic pressure constraints for the aircraft are given by

$$h \ge 0 \quad (9)$$

$$q \le q_{max} \quad (10)$$

For computational purposes we introduce the "vertical and horizontal load factor components":

$$n_v := n \cos\mu \tag{11}$$

$$n_h := n \sin\mu \tag{12}$$

Using an appropriate subscript i = A1,A2,M1,M2 to identify the vehicle, the state and control variables of vehicle i are given by

$$z_i^T := (x_i, y_i, h_i, V_i, \gamma_i, \chi_i) \tag{13}$$

$$u_i^T := (\xi_i, n_{hi}, n_{vi}) \tag{14}$$

During the pre-launch maneuvering phase, each aircraft controls the joint state

$$z^T := (z_{A1}^T, z_{A2}^T) \tag{15}$$

in order to obtain a successful launch opportunity to itself and to deny the same to the opponent if possible.

The two essential criteria associated with any joint state are the predicted miss distances $R_{12}(z)$ and $R_{21}(z)$. They indicate the result of a (fictitious) missile firing of A1 against A2 and vice versa, respectively (see Fig.1). $R_{ij}(z)$ denotes the final distance in the pursuit-evasion encounter between M_i and A_j. It can be obtained in a missile/target simulation starting at z and terminating at a prescribed value of the closing speed between M_i and A_j. $R_{ij}(z)$ is well defined since the missile and aircraft guidance laws are fixed in the post-launch phase. The missile is guided by pure proportional navigation. The evasion strategy of the aircraft is basically taken from Ref.9. It has been extended by auxiliary control modes to comply with the constraints (9) and (10) (see Ref.8). Inherently, it is assumed that the missiles are autonomous after launch and immediately detected by the respective target aircraft. The missile models are only needed to evaluate $R_{ij}(z)$.

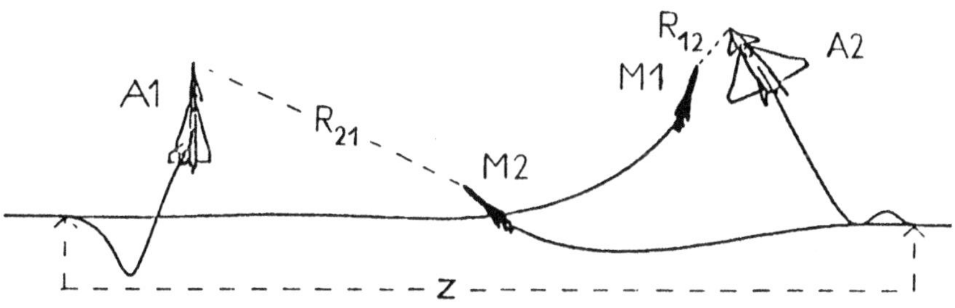

<u>Figure 1:</u> The predicted miss distances R_{12} and R_{21} in a missile exchange between two aircraft

Denoting d_i as the maximum miss distance for M_i to score a hit, we consider that the engagement starts at an inital state z_o such that neither aircraft can hit the opponent with its missile, i.e.,

$$R_{12}(z_o) > d_1, \; R_{21}(z_o) > d_2 \tag{16}$$

We study the guidance problem of A1 in the sequel. A1 is assumed to have complete and perfect state and model information and precise knowledge of the missile and aircraft guidance laws for the post-launch phase. Furthermore, A2 is assumed first-acting and A1 second-acting in a "response" mode (Ref.7) at every instant of time. I.e. u_{A2} is known to A1.

3. <u>Design of the Guidance Law</u>

A1 tries to achieve a safe launch position, i.e.

$$R_{12}(z) = d_1 \;, \; R_{21}(z) > d_2 \tag{17}$$

Classical game approaches to optimize the guidance of A1 consider the whole time interval up to the termination condition (17). However, since A2's future strategy is unknown, the present approach suggests maneuvering on an incremental basis. The objective is to deteriorate

A2's missile launch condition as much as possible subject to a certain stipulated improvement ϵ of own launch condition. This requires the solution of the following problem at every instant t:

$$\underset{u_{A1}}{\text{minimize}}\ (dR_{21}/dt)/\ (dR_{12}/dt) \qquad \text{(ratio criterion)} \qquad (18)$$

$$\text{subject to } (dR_{12}/dt) \leq -\epsilon < 0$$

u_{A2} occurs explicitly in the rates above. Thus (18) can be regarded as "response" strategy to exploit nonoptimal play by A2 (Ref.7). Strategy (18) obtained on an incremental basis is an approximation of the minimum risk capture strategy for A1 which attempts to achieve the successful launch condition $R_{12}(z) = d_1$ with as large $R_{21}(z)$ as possible. The exact strategy involves the complete global solution of a combat game with the target sets $R_{12} \leq d_1$ and $R_{21} \leq d_2$. It is not feasible with present computational methods.

4. Simulation Program for Performance Evaluation

A simulation program has been developed for the performance evaluation of the proposed response strategy based on the ratio criterion (18). These simulations were done within the frame of the general simulation program MISI (Ref.10). The F-4 aircraft model is chosen for both A1 and A2. R_{12} and R_{21} are evaluated at a final closing speed of 150 m/sec. Further constants are $d_1 = d_2 = 1$ km and $\epsilon = 100$ m/sec. A2 is guided by proportional navigation; lift and thrust commands for A1 are generated by the proposed scheme. Fourth order Runge-Kutta method with a fixed step size of 0.1 sec is used to integrate the vehicle equations of motion. The simulation is continued until A1 or A2 obtains a successful launch opportunity. The terrain and dynamic pressure constraints (9), (10) for A1 are indirectly satisfied by adding further control constraints to the optimization problem (18):

$$n_v \geq n_{vmin}(h,V,\gamma) \qquad (19)$$

$$\dot{q} \leq \dot{q}_{ref} = C(q_{max}-q). \qquad (20)$$

n_{vmin} is the lift command to perform a specified ground avoidance maneuver. Constraint (20), which becomes active already for $q < q_{max}$, is to avoid an abrupt entry on to the q_{max}-boundary. Analytical optimization of (18) subject to the control constraints (7), (8), (19), (20) is possible if a parabolic drag polar is assumed. It shows that partial throttle is possible only for the case $\dot{R}_{12} = -\epsilon$ and $n = n_{max}$. For a general model of the drag polar the optimization (18) must be done numerically. The analytical solution for a parabolic approximation provides an excellent start value, which is essential for convergence of the numerical method. The reason is the proximity of the singularity $\dot{R}_{12} = 0$, which generally crosses the control set bounded by (7), (8).

For each evaluation of u_{A1} gradient components of $R_{ij}(z)$ are needed. Altogether 10 missile target/simulations are necessary to provide the required partials for (18) approximated as finite differences. Because of the computational effort these partials are held constant within an "update interval" of 0.5 sec.

5. Results and Discussion

For all examples mentioned in this section the flight paths of A1 are depicted in the altitude/Machnumber diagram Fig.7. Common initial values of all cases are $x_{A1} = y_{A1} = y_{A2} = 0$.

Fig.2 shows the results for the symmetric initial condition

$$h_{A1} = 7\text{ km},\ V_{A1} = 437\text{ m/sec},\ \gamma_{A1} = 0^\circ,\ \chi_{A1} = 15^\circ \qquad (21)$$
$$h_{A2} = 7\text{ km},\ V_{A2} = 437\text{ m/sec},\ \gamma_{A2} = 0^\circ,\ \chi_{A2} = 165^\circ$$
$$x_{A2} = 30\text{ km}$$

The first seconds of A1's trajectory show the transition to an orientation which is a compromise between collision course and an evasion direction corresponding to the attack direction of A2. High control effort is commanded and even the throttle is cut off resulting in a loss of speed. This increases R_{21} while keeping $\dot{R}_{12} = -\epsilon$; the relative advantage $R_{21}-R_{12}$ builds up to 1 km. This flight phase is followed by one in which the control effort is low. Here $R_{21}-R_{12}$ improves more slowly until a successful launch condition is achieved for A1 with a safety margin of $R_{21}-R_{12}$ = 1.67 km. Note that the piecewise constance of R_{12} and R_{21} is due to the fact that both functions are only updated twice a second.

Simulation runs are repeated for the same initial conditions but with different values of ϵ. The results do not depend much on ϵ.

Next, simulations are carried out changing only A1's initial heading. All those runs terminated with nearly the same values of γ_{A1} and χ_{A1} as in Fig.2. This suggests the validity of quasisteady limit values defined by the condition that the minimum of (18) is attained at $n_{v_{A1}} = 1$, $n_{h_{A1}} = 0$, and $\xi_{A1} = 1$. If the initial orientation of A1 is close to these values the advantage in $R_{21} - R_{12}$ is immediate (Fig.3: $\chi_{A1}(0) = 50^o \approx \chi_{A1}(t_f)$). Fig.4 shows the performance plot for the initial condition

$$\begin{aligned} h_{A1} &= 8 \text{ km}, V_{A1} = 417 \text{ m/sec}, \gamma_{A1} = 0^o, \chi_{A1} = 23^o \\ h_{A2} &= 11 \text{ km}, V_{A2} = 413 \text{ m/sec}, \gamma_{A2} = 0^o, \chi_{A2} = 165^o \\ x_{A2} &= 35 \text{ km} \end{aligned} \tag{22}$$

Although A1 looses speed in the climb still $R_{21} - R_{12}$ changes sign and A1 gains a successful launch condition in the end. Fig.5 shows the performance plot when A1 has an altitude advantage:

$$\begin{aligned} h_{A1} &= 11 \text{ km}, V_{A1} = 472 \text{ m/sec}, \gamma_{A1} = 0^o, \chi_{A1} = 100^o \\ h_{A2} &= 8 \text{ km}, V_{A2} = 304{,}5 \text{ m/sec}, \gamma_{A2} = 20^o, \chi_{A2} = 165^o \\ x_{A2} &= 25 \text{ km} \end{aligned} \tag{23}$$

Here R_{21} increases in the second half building up $R_{21} - R_{12}$ briskly to 5 km. Then the method hangs up because of the approaching q_{max}-boundary: The inequalities $\dot{R}_{12} \leq -\epsilon$ and (20) have no common solutions at the end. A possible remedy would be to minimize $\dot{R}_{12}$ alone considering the relative advantage of 5 km in A1's favour. Fig.6 shows the results for the same symmetric initial conditions (21) as in Fig.2; but now A2 switches to avoidance when R_{12} shrinks to 3 km while A1 continues playing strategy (18). R_{21} reduces dramatically relative to R_{12} tilting the advantage in A2's favour who gains a successful launch condition.

A1 might have done better by switching to a survival strategy when the rapid decrease of R_{21} commences. In summary, the examples indicate that the guidance law should be augmented by control modes for survival and pursuit.

6. Conclusions

The feedback guidance for pre-launch maneuvering presented in this paper takes advantage of opponents' errors in strategy at any time by a quick rataliatory mode of play. Simulation study shows that the scheme orients the initial attack direction of the aircraft as a compromise between its capture and survival objectives, although at a high control effort. However, subsequent corrections to this direction require relatively low values of load factor. The performance of the guidance scheme can be improved further by incorporating features of threat reciprocity. This seems required only for rather large and small values of the security margin. A major disadvantage of the method is its high computational expense. The missile/target simulations needed for generating the control actions have to be speeded up considerably to have a genuine real-time algorithm.

Acknowledgements

The authors are grateful to Dr. K.H. Well for the discussions they had with him while this work was in progress. They also would like to thank Mrs. S. Hagen for her help with the simulation program, and Mrs. G. Rossmeier for her excellent typing of the manuscript.

References

[1] Herbst, W., "Dynamics of Air Combat", Journal of Aircraft, Vol.20, pp.594-598, 1983.

[2] Järmark, B., "A Missile Duel Between Two Aircraft", Journal of Guidance and Control, Vol.8, pp.508-513, 1985.

[3] Moritz, K., Polis, R., and Well, K.H., "Pursuit-Evasion in Medium-Range Air-Combat Scenarios", Comp. Math. Appl. Vol.13, No.1/3, pp.167-180, 1987.

[4] Shinar, J., and Davidovitz, A., "A Qualitative Analysis of Future
Air Combat with 'Fire and Forget' Missiles", AIAA paper No.87-2315, AIAA Guidance, Navigation and Control Conference, Monterey, Ca., USA, August 17-19, 1987.

[5] Kelley, H.J., "A Threat-Reciprocity Concept for Pursuit-Evasion" Differential Games and Control Theory II, edited by E.O. Roxin, P.T. Liu, and R.L. Sternberg, Marcel Dekker, Inc., New York, 1977.

[6] Kelley, H.J., Cliff, E.M., and Lefton, L., "Reprisal Strategies in Pursuit Games", Journal of Guidance and Control, Vol.3, pp.257-260, 1980.

[7] Prasad, U.R., Ghose, D., "Formulation and Analysis of Combat Problems as Zero-Sum Bicriterion Differential Games", to appear in JOTA, 1988.

[8] Grimm, W., "Missile Avoidance Simulations With a Guidance Law by J.Shinar", Internal Report No. 515-85/7, DFVLR Oberpfaffenhofen, FRG, 1985.

[9] Shinar, J., and Gazit, R., "Optimal 'No-Escape' Firing Envelopes of Guided Missiles", AIAA paper No.85-1960, AIAA Guidance, Navigation and Control Conference, Snowmass, Co., USA, August 1985.

[10] Somieski, G., "MISI Version 2 - Modular Interactive Simulation - Installation and Operation Manual", DFVLR Report No.87-02, DFVLR Oberpfaffenhofen, FRG, 1987.

__Figs.2a-d:__ Simulation results with initial state (21).
A2: PN-guidance throughout

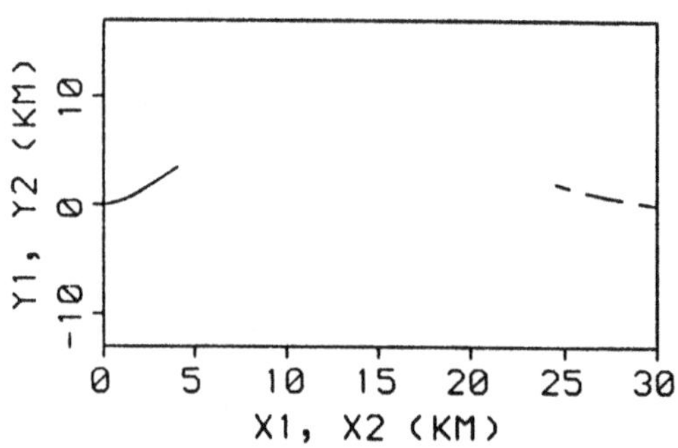

__Fig.2a:__ Top view of the flight paths. A1=solid line, A2=dashed line
X1 = x_{A1} etc.

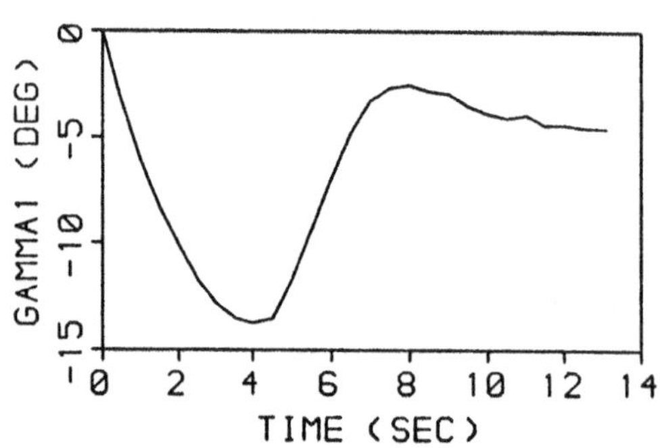

__Fig.2b:__ GAMMA1 $\hat{=}$ γ_{A1}

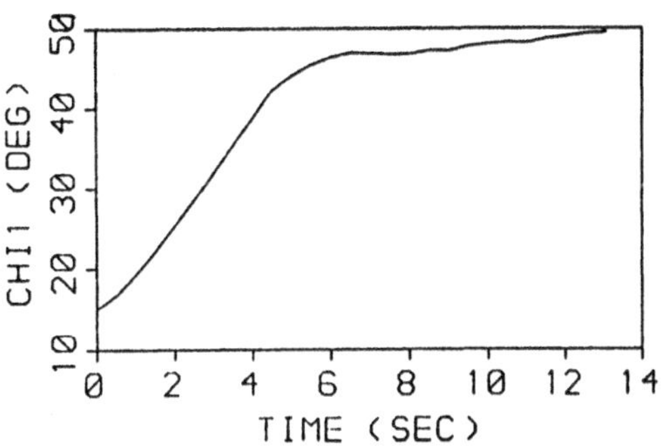

Fig.2c: CHI1 $\hat{=} \chi_{A1}$

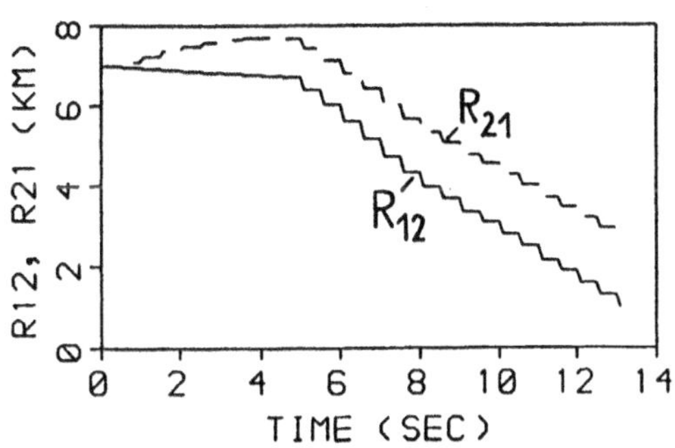

Fig.2d: R_{12} = solid line
R_{21} = dashed line

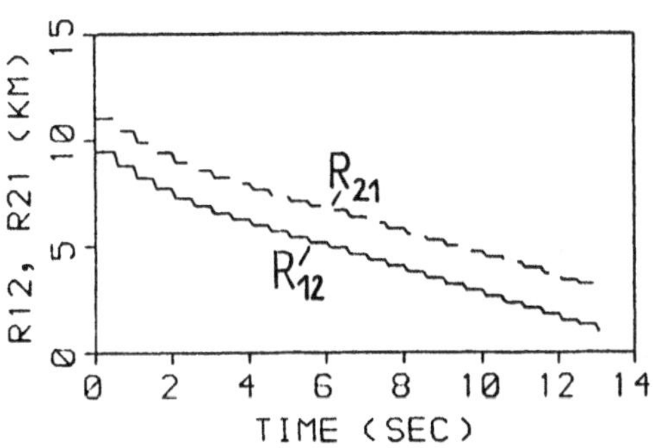

Fig.3: R_{12} and R_{21} for initial state (21) with $\chi_{A1}(0) = 50^{\circ}$

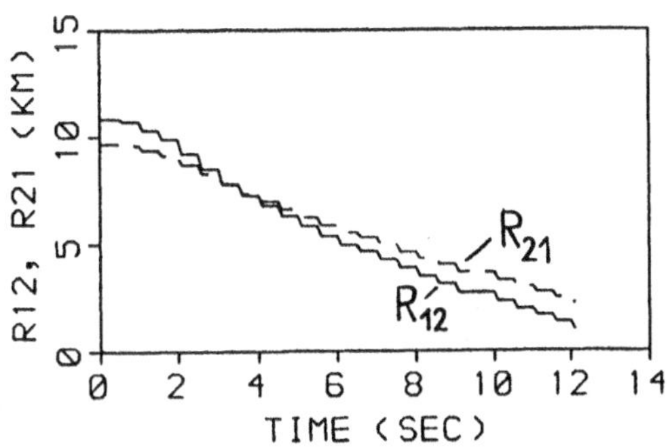

Fig.4: R_{12} and R_{21} for initial state (22)

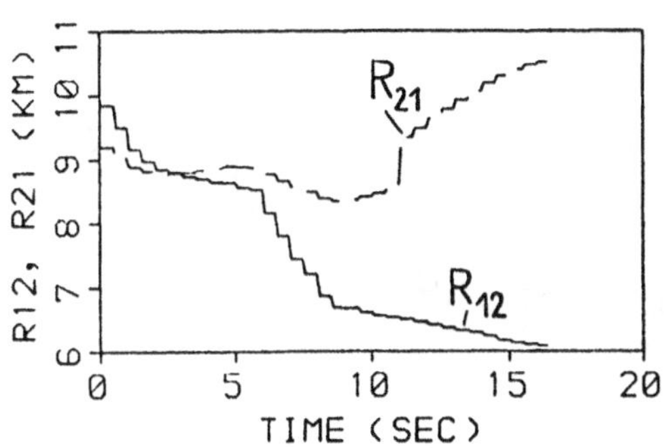

Fig.5: R_{12} and R_{21} for initial state (23)

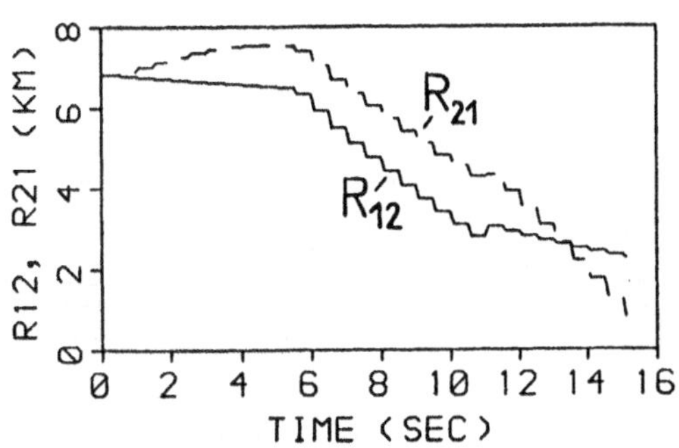

Fig.6: R_{12} and R_{21} for initial state (21), A2 evades for $R_{12} \leq 3$ km.

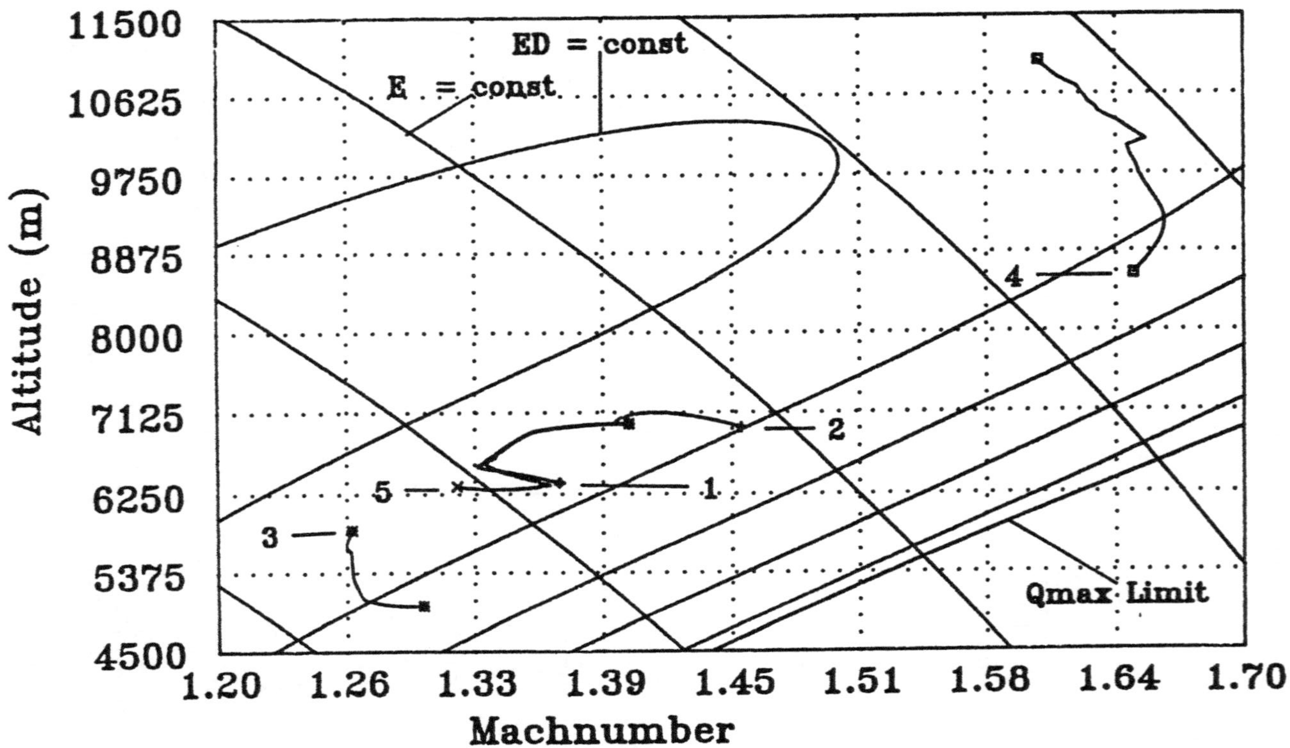

<u>Fig.7:</u> A1-trajectories in the altitude/Machnumber diagram

1: initial state (21), PN-guidance for A2 (see also Figs.2a-d).

2: like 1, but $\chi_{A1}(0) = 50^{\circ}$ (see also Fig.3).

3: inital state (22) (see also Fig.4).

4: inital state (23) (see also Fig.5).

5: like 1, but A2 evades for $R_{12} \lesssim 3$ km (see also Fig.6).

GUIDANCE LAW SYNTHESIS BASED ON A PLANAR PURSUIT-EVASION GAME SOLUTION

A. Green[1], J. Shinar[2], M. Guelman[3]

ABSTRACT

Closed form solution of a coplanar pursuit-evasion game between a coasting missile and a maneuvering target of constant speed leads to express the optimal strategy of the missile as a function of the current and final values of the state variables. Based on this solution a feedback guidance law for a real time implementation is synthesized. The performance achieved by this guidance law is substantially superior to proportional navigation and only slightly worse than the optimal game strategy.

1. INTRODUCTION

Realistic pursuit-evasion engagements opposing a missile to a maneuvering target in the atmosphere are characterized by an inherent dissymmetry. The missile has a complete advantage in speed and maneuverability over the aircraft. This kinematical advantage is, however, temporary. The rocket motor accelerating the missile to a high velocity, is of short duration. In the coasting phase the high, but finite kinetic energy of the missile is continuously dissipated by the work done against the aerodynamic drag. On the other hand the aircraft energy resources (fuel) can be considered, compared to the missile almost unlimited. During the relatively short length of a missile fly-out (20-50 sec.) the aircraft can keep its velocity almost constant. This asymmetry results in the well known limits of missile's firing envelopes.

The purpose of the present paper is to synthesize a feedback guidance law for a missile. This feedback guidance law is based on the closed form

---

[1]Graduate student, Faculty of Aeronautical Engineering, Technion - Israel Institute of Technology, Haifa, Israel

[2]Professor, Faculty of Aeronautical Engineering - Technion - Israel Institute of Technology, Haifa, Israel, Associate Fellow AIAA

[3]Deputy Director, Guidance Division, RAFAEL, M.O.D., P.O. Box 2250, Haifa, Israel.

solution of a planar pursuit-evasion game between a coasting pursuer and a maneuvering evader of contant speed [1].

2. PROBLEM DEFINITION

The geometry of planar pursuit defining the state variables of the game is depicted in Fig. 1. A vehicle P, called the pursuer, possessing a velocity V_P is pursuing in planar motion a second vehicle, the evader E, assumed to be flying with a constant velocity V_E.

The kinematic equations are

$$\dot{R} = V_E\cos(\gamma_E-\theta) - V_P\cos(\gamma_P-\theta) \quad (1)$$

$$\dot{\theta} = [V_E\sin(\gamma_E-\theta)-V_P\sin(\gamma_P-\theta)]/R \quad (2)$$

$$\dot{\gamma}_E = \Gamma_E u_E \quad ; \quad |u_E|\leq 1 \quad (3)$$

$$\dot{\gamma}_P = \Gamma_P u_P \quad ; \quad |u_P|\leq 1 \quad (4)$$

where Γ_E and Γ_P are the maximum turning rates of the evader and the pursuer respectively and u_E and u_P are the controls. These values depend on the maximum lateral acceleration available to the players and are therefore functions of the flight conditions. For a constant speed vehicle in horizontal flight the value of Γ is constant.

The pursuer speed dynamics is given by,

$$\dot{V}_P = -k_P V_P^2 C_D \quad (5)$$

where k_P is a constant depending on the altitude and C_D is the nondimensional drag coefficient. Neglecting gravity, the turning rate of the pursuer is given by,

$$\Gamma_P u_P = k_P V_P C_L \quad (6)$$

where C_L is the nondimensional lift coefficient, having a limit value of $C_{L_{max}}$.

For missiles operating in the domain of small angles of attack, a linear lift

$$C_L = C_{L_\alpha}\alpha \tag{7}$$

and a parabolic drag polar

$$C_D = C_{D_o} + (KC_{L_\alpha}^2)\alpha^2 \tag{8}$$

models are appropriate.

One can define new normalized variables,

$$r = R/R_{ref} \tag{9}$$

$$v = V_P/V_E \tag{10}$$

$$t' = tV_E/R_{ref} \tag{11}$$

where R_{ref} is the minimum admissible turning radius of the pursuer defined by,

$$R_{ref} \triangleq V_P/\Gamma_P = 1/[k_P C_{L_{max}}] \tag{12}$$

With these new variables, and with a dot denoting now the derivative with respect to normalized time t' the game equations with the parabolic drag polar are given by,

$$\dot{r} = \cos(\gamma_E-\theta) - v\cos(\gamma_P-\theta) \tag{13}$$

$$\dot{\theta} = [\sin(\gamma_E-\theta) - v\sin(\gamma_P-\theta)]/r \tag{14}$$

$$\dot{\gamma}_E = \sigma u_E \tag{15}$$

$$\dot{v} = -v^2(a+bu_P^2) \tag{16}$$

$$\dot{\gamma}_P = vu_P \tag{17}$$

where σ is the ratio of pursuer's minimum turning radius to that of the evader (in general $\sigma<1$) and

$$a = C_{D_o}/C_{L_{max}} \tag{18}$$

$$b = KC_{L_{max}} \tag{19}$$

The game described by these equations terminates with capture when the pursuer approaches the evader to the normalized distance

$$r = r_f \tag{20}$$

i.e., the game target set T is defined as a closed circular cylinder of radius r_f,

$$T = \{x \in E^5 : r \leq r_f\}, \tag{21}$$

where $x=(r,\theta,\gamma_E,v,\gamma_P)^T$ is the state vector and no additional conditions are imposed on θ, γ_E, v, γ_P.

3. DERIVATION OF OPTIMAL STRATEGIES

The solution of a game of kind, such as one formulated in the previous section of this paper, is to determine the boundary of the "capture set" in the game. This "capture set" is the set of all admissible initial conditions $x_o=(r_o,\theta_o,\gamma_{Eo},v_o,\gamma_{Po})^T$ from which the pursuer can drive the game to the target set T given in (21) with $\dot{r}_f<0$, against any admissible control action of the evader.

As discussed in the Introduction, the "capture set" of the present game is a bounded region in the 5 dimensional state space. The boundary of the "capture set" consists of the target set T, the hyperplane $v=v_o$, (determined by the initial kinetic energy of the pursuer) and a closed semipermeable hypersurface called the "barrier". The "barrier" is formed by an infinite set of game trajectories [solutions of Eqs. (13)-(17)] generated by using a pair of optimal strategies $E^*(x)$ and $P^*(x)$ such that

$$u_E^*(t') = E^*[x(t')] \tag{22}$$

and

$$u_P^*(t') = P^*[x(t')] \tag{23}$$

The solution of this game is presented in Ref. [1].

The evader strategy is a hard turn towards the final line of sight direction

$$u_E^*(t') = -\mathrm{sign}[\sin(\gamma_E-\theta_f)] \tag{24}$$

This strategy is identical to the one obtained in the "game of two cars" [2].

The pursuer strategy is defined by

$$u_P^*(t') = \frac{P_{\gamma_E} - r\sin(\theta-\theta_f)}{2b\cos(\gamma_{E_f}-\theta_f)\tau} \tag{25}$$

where

$$P_{\gamma_E} = -\frac{1}{\sigma}[\cos(\gamma_{E_f}-\theta_f) - \cos(\gamma_E-\theta_f)]\mathrm{sign}[\sin(\gamma_E-\theta_f)] \tag{26}$$

and

$$\tau = t_f' - t' \tag{27}$$

is the normalized time to go that can be explicitly expressed as a function of the state variables and their final values in the form of

$$\tau = A + (A^2+B^2)^{1/2} \tag{28}$$

with

$$A = \frac{1}{2a}\left[\frac{\cos(\gamma_P-\theta_f)}{\cos(\gamma_{E_f}-\theta_f)} - \frac{1}{v}\right] \tag{29}$$

and

$$B = \frac{P_{\gamma_P}}{(2ab)^{1/2}\cos(\gamma_{E_f}-\theta_f)} \tag{30}$$

where

$$P_{\gamma_P} = r\sin(\theta-\theta_f) - P_{\gamma_E} \tag{31}$$

At the end of the game the pursuer's optimal control is

$$u_P^*(t_f') = - tg(\gamma_{P_f}-\theta_f)/2b \tag{32}$$

This expression indicates that the pursuer's strategy is to turn toward the final line of sight direction but this direction is never reached. Such behavior is similar to the one observed in [3].

Since the optimal control strategies of the players are expressed by explicit functions of the state variables and their final values, barrier trajectories can be constructed by backwards integration starting at the boundary of the usable part (BUP) of the target set.

4. BARRIER CONSTRUCTION

For the barrier construction the four dimensional reduced space r, ϕ_E, ϕ_P, v is employed with

$$\phi_i = \gamma_i - \theta \qquad i = E, P \tag{33}$$

without loosing any generality, the final line of sight was employed as the angular reference line, i.e., $\theta_f = 0$.

In the reduced space the target set boundary as well as the barrier, are three dimensional manifolds. The barrier is supported by the two dimensional BUP, defined by

$$r = r_f, \quad v_f = \cos\gamma_{E_f}/\cos\gamma_{P_f} \tag{34}$$

An example of the game with the set of parameters, detailed in Table 1, was solved. In this example the existence of a "dispersal surface" on the barrier was discovered by the intersection of two families of barrier trajectories [1].

On the "dispersal surface" both players have two different strategies. The choice between the two strategies (left or right turn) is the evader's and the pursuer only reacts to this choice. In Fig. 2 the intersection of the capture zone with $v=v_o$ is shown, including the dispersal line.

The intersection of this barrier with the hyperplane $v=v_o$=const is a two dimensional surface characterized by the set of points (r_o, ϕ_{E_o}, ϕ_{P_o}). For a given value of v_o the "maximum capture range" r_o is a function of the initial angular geometry, represented by ϕ_{E_o} and ϕ_{P_o}. This relationship is plotted in Fig. 3 for the presently used example.

Table 1 Parameters of the game (Parabolic drag polar).

Parameter	Symbol	Value
Minimum admissible turning radius of the pursuer.	R_{ref}	1515.15 [m]
Evader Velocity	V_E	300 [m/sec]
Ratio of pursuer's minimum turning radius to that of the evader	σ	0.809
Ratio of parasitic drag to maximum lift	a	0.0875
Ratio of induced drag to lift at maximum lift coefficient	b	0.40

5. COMPARISON WITH PROPORTIONAL NAVIGATION

In order to evaluate the potential benefit which can be derived from the qualitative game solution, a comparison of the pursuer barrier strategy with proportional navigation (PN), the most popular missile guidance law, was carried out.

The results show that the improvement in increased capture range provided by the optimal pursuer strategy is substantial in all cases. The exact level of improvement depends, of course, on the initial conditions and the navigation ratio (N) of the PN guidance law selected for the comparison. For initial conditions favorable to the missile, i.e. near to collision course, the increase in capture range is only of the order of 3-5%. For unfavorable initial conditions, however, the improvement can reach 15-20%. This trend is demonstrated by the results of three examples, presented in Table 2 in comparison with a PN missile of N=4. (This value is asserted to be the lowest one which provides an efficient guidance against maneuvering targets [4] in a scenario of nonlinear geometry.)

Table 2 Comparison with a PN missile

No	ϕ_{P_o}	ϕ_{E_o}	$u^*_{E_o}$	R^*_o	$R_o(PN)$	$\frac{R^*_o - R_o(PN)}{R^*_o(PN)}$
1	20°	100°	+1(right)	7289m	7026m	3.74%
2	20°	170.3°	+1(right)	10797m	10169	6.18%
3	20°	260°	-1(left)	6061m	5122	18.33%

Note that against the PN missile the evader's optimal game strategy is not the very best. Against such a missile a "bang-bang" strategy [5] will further decrease the capture range.

The demonstrated advantage of the game solution suggests an eventual implementation of the pursuer's optimal strategy. For this purpose this strategy has to be expressed in an explicit feedback form, or at least approximated by feedback expressions.

6. GUIDANCE LAW SYNTHESIS

The optimal strategy of the pursuer, given in (25) and making use of (26), depends on 3 quantities, $(\tau, \theta_f, \gamma_{E_f})$, unknown at any current position. Since τ can be expressed as a function of the current and final state, the number of unknowns is reduced to two (θ_f, γ_{E_f}).

Observing the nature of the optimal evader strategy, it can be assumed that the value of $(\gamma_{E_f}-\theta_f)$ in most cases will be very small, satisfying

$$\cos(\gamma_{E_f}-\theta_f) \approx 1 \tag{35}$$

Based on this approximation, only the value of θ_f has to be computed (or estimated), for expressing the pursuer's guidance law in the form of

$$\tilde{U}_P = \tilde{U}_P\,(v,\phi_P,\phi_E,\tilde{\theta}_f) \tag{36}$$

In the present paper the values of the final line of sight direction were obtained from the game solution as a function of the intial conditions

$$\tilde{\theta}_f = \tilde{\theta}_f(v_o,\phi_{P_o},\phi_{E_o}) \tag{37}$$

and approximated in a polynomial form

$$\tilde{\theta}_f = \sum_{i=o}^{n} a_i(v_o,\phi_{P_o})\phi_{E_o}^{i} \tag{38}$$

where a_i are coefficients depending on the initial pursuer's to evader's velocity ratio v_o and the initial pursuer's look angle ϕ_{P_o}. $\tilde{\theta}_f$ have been computed with n=4 and the results have an excellent accuracy of $\pm$ 0.05°.

On the dispersal line, for the same set of initial conditions, the value of θ_f depends on the direction of the evader turn. Since for the evader both turns are equivalent, its decision cannot be predicted. If in the pursuer guidance law the wrong value of θ_f is used, poor results are obtained. Therefore on the dispersal line the pursuer has to wait until the direction of the evader turn is identified and the optimal game strategy with the correct value of θ_f can be implemented. Because of that waiting period the actual capture range will be smaller than the game optimal solution, but it will approach asymptotically this value as the waiting period approaches zero. During that period, required for identifying the evader's turning direction, some other guidance law, such as PN, can be used. For a set of initial conditions on the dispersal line (r_o=7.126, v_o=2.67, ϕ_{P_o}=20°, ϕ_{E_o}=170.3°), the line of sight rate is depicted in Fig. 4 assuming a pursuer guided by PN with N=4. It can be seen that the line of sight rate behavior is clearly different for the two optimal turning options of the evader. Therefore the line of sight rate can be used as a criterion to distinguish between right and left evader turns.

Based on the above discussion a feedback guidance law for the pursuer is proposed. This guidance law has to be robust for any possible (optimal and non-optimal evader maneuver.)

The proposed guidance law is the synthesis of the following elements:

a. At t=0 start guidance using PN during a fixed period Δt.

b. During this period, compute:

1) Based on the state variables $(r_o, v_o, \phi_{P_o}, \phi_{E_o})$ and the game solution, the final values of the line of sight direction assuming the evader turns either to the right $(\tilde{\theta}_f)_R$ or to the left $(\tilde{\theta}_f)_L$.

2) The line of sight rates $\dot{\theta}_R$, $\dot{\theta}_L$ obtained by the end of that period $(t=\Delta t)$ by each maximum evader turn.

c) Measure $\dot{\theta}$ at $t=\Delta t$.

1) If $\min\{|\dot{\theta}-\dot{\theta}_R|, |\dot{\theta}-\dot{\theta}_L|\} > \varepsilon$ continue with PN for the next time step and go to a.

2) If $|\dot{\theta}-\dot{\theta}_R| \leq \varepsilon$ switch to optimal game strategy (36) with $\theta_f=(\tilde{\theta}_f)_R$.

3) If $|\dot{\theta}-\dot{\theta}_L| \leq \varepsilon$ switch to optimal game strategy (36) with $\theta_f=(\tilde{\theta}_f)_L$.

d. Compute the expected time history of the line of sight rate $\tilde{\dot{\theta}}$ along the optimal trajectory and compare it to the actually measured value of $\dot{\theta}$ at the end of each time step.

e. If $|\dot{\theta}-\dot{\theta}| > \varepsilon$ switch to PN guidance and go back to a.

This guidance law approximates the optimal game solution for optimal evader strategy and guarantees robustness against a non-optimal evader behavior.

The open loop nature of the game strategy and the estimation of $\tilde{\theta}_f$ even with a very small error will cause a miss distance (parallel pursuer and evader trajectories). Therefore at the terminal phase a different implementation is needed.

Observing the nature of the optimal pursuer strategy it can be assumed that the valaue of ϕ_P at the final phase of the engagement will be small, satisfying

$$tg\ \phi_P \approx \phi_P \tag{39}$$

Based on this assumption and making use of (32), the pursuer strategy in the neighborhood of the target set can be approximated in the form

$$u_P^*(\tau) \approx -\frac{1}{2b}(\phi_P + \tau' \dot{\phi}_P) \tag{40}$$

This is a lead pursuit guidance with a lead angle proportional to the time-to-go, which is a function of the state variables as shown in Eqs. (28)-(31).

The final phase of the pursuit can be defined by

$$v < v_{lim} \tag{41}$$

The suggested terminal guidance law is

1. If $|\dot{\theta}-\dot{\tilde{\theta}}| \leq \varepsilon$ we use lead pursuit (40).
2. If $|\dot{\theta}-\dot{\tilde{\theta}}| > \varepsilon$ use PN guidance.

The proposed implementation of the feedback guidance law is summarized in the flow chart in Fig. 5.

The value of ε and v_{lim} are selected in order to guarantee a robust guidance performance.

Obviously the maximum range of the proposed (suboptimal) guidance law, R_o, will be shorter than the capture range of the game solution.

7. SENSITIVITY TESTS

For the initial conditions of the three examples in Table 2, $\tilde{R}_o$, the maximum range achieved with the proposed guidance law, (with v_{lim}=1.5, ε=2.5x10^{-3} and $\Delta t \to 0$) was compared both with R_o^*, the maximum game range, and R_o(PN) the maximum PN range, against an optimal evader strategy. The results summarized in Table 3 show that the loss in performance of the proposed guidance law due to the approximate terminal guidance is truly negligible compared to the improvement achieved with respect to PN.

Table 3. Maximum range of the proposed guidance law ($\Delta t=0$)

No	ϕ_{P_o}	ϕ_{E_o}	R_o^* [m]	t_f^*[sec]	R_o(PN) [m]	$\tilde{R}_o$ [m]	$\frac{\tilde{R}_o-R_o^*}{R_o^*}$	$\frac{\tilde{R}_o-R_o(PN)}{R_o(PN)}$
1	20°	100°	7289	36.0	7026	7288	-0.014%	3.73 %
2	20°	170.3°	10797	36.1	10169	10795	-0.019%	6.16 %
3	20°	260°	6061	33.1	5122	6060	-0.017%	18.31 %

For the initial conditions of example 2 (on the dispersal line), $\tilde{R}_o$ was compared with R_o^* and R_o(PN) using again v_{lim}=1.5 and ε=2.5x10^{-3} but with $\Delta t \neq 0$ against an optimal evader strategy. The results presented in Table 4 show that the pursuer can wait as long as 1-2 seconds to identify the evader turning direction for estimating $\tilde{\theta}_f$ without a substantial loss of capture range.

Table 4. Maximum range of the proposed guidance law ($\Delta t \neq 0$).

Δt[sec]	$\tilde{R}_o$[m]	$\frac{\tilde{R}_o-R_o^*}{R_o^*}$	$\frac{\tilde{R}_o-R_o(PN)}{R_o^*(PN)}$
0.0	10795	-0.019%	6.16 %
0.5	10769	-0.259%	5.90 %
1.0	10747	-0.463%	5.68 %
1.5	10723	-0.685%	5.45 %
2.0	10698	-0.917%	5.20 %

In the present game solution the final closing velocity is zero, as it can be seen in Eq. (34). In real missile/aircraft engagements a final closing velocity of the order of 150 m/sec is requested for effective operation. Though such a constraint reduces the maximum range, it has to be taken into account for any realistic evaluation. Therefore, the results of Table 3 are recomputed for $v_{c_f}=0.5$ and $v_{lim}=2.0$ and presented in Table 5. The essential improvement induced by the optimal guidance law implementation does not change by taking into account a realistic constraint on the final closing velocity. A large set of test examples demonstrated that the proposed guidance law implementation is robust against all possible evader maneuvers.

Table 5. Reduction in maximum range due to $v_{c_f} \neq 0$ ($\Delta t=0$, $v_{c_f}=0.5$)

No	ϕ_{P_o}	ϕ_{E_o}	R_o^* [m]	t_f^*[sec]	$\Delta\hat{R}_o$ [m]	Δt_f[sec]
1	20°	100°	5950	16.6	-1338	-19.4
2	20°	170.3°	9400	16.7	-1395	-19.4
3	20°	260°	3900	12.3	-2160	-20.8

8. CONCLUSIONS

In the present paper a missile guidance law based on the qualitative solution of the planar pursuit-evasion game between a coasting missile and a constant speed aircraft is outlined. This guidance law provides a substantial improvement by enlarging the capture zone in comparison with a missile guided by proportional navigation. A real-time feedback implementation of this guidance law is proposed, taking into account realistic constraints on the final closing velocity. It is shown to be robust with respect to non-optimal evader strategies. The loss of performance of this suboptimal feedback guidance law depends on the time period necessary to identify the evader's actual maneuver, but it is rather small compared to the optimal game solution.

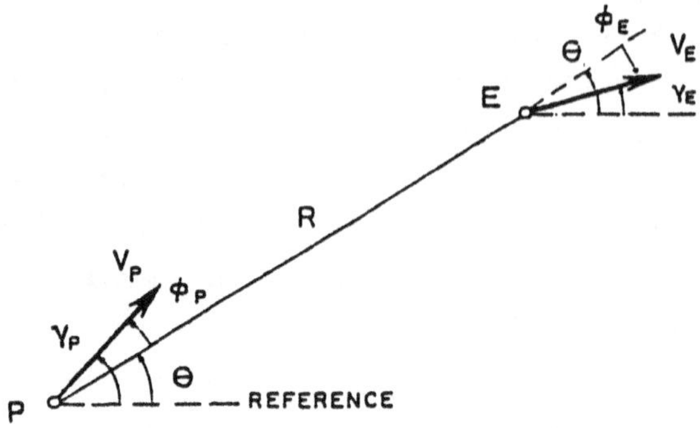

Fig. 1 Geometry of planar pursuit.

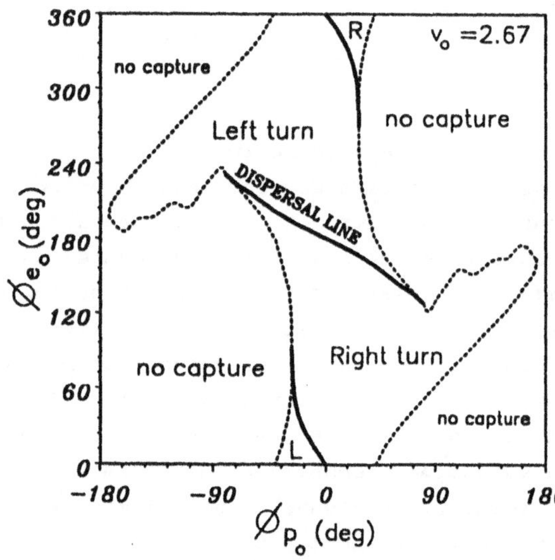

Fig. 2 Intersection of the capture zone with $v=v_o$, (Projection on the ϕ_P-ϕ_E plane).

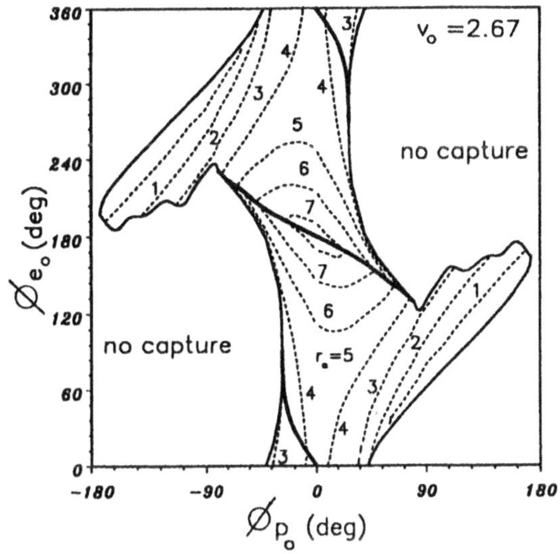

Fig. 3 Map of maximum capture range at $v=v_o$ (view from infinite r).

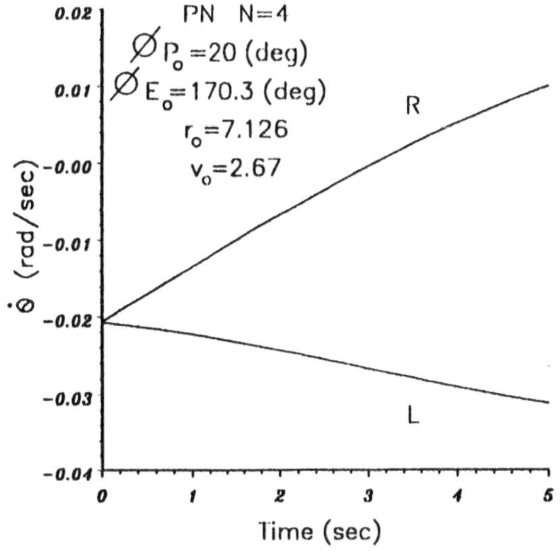

Fig. 4 Line of sight rate for a PN pursuer with N=4 (initial conditions on the dispersal line).

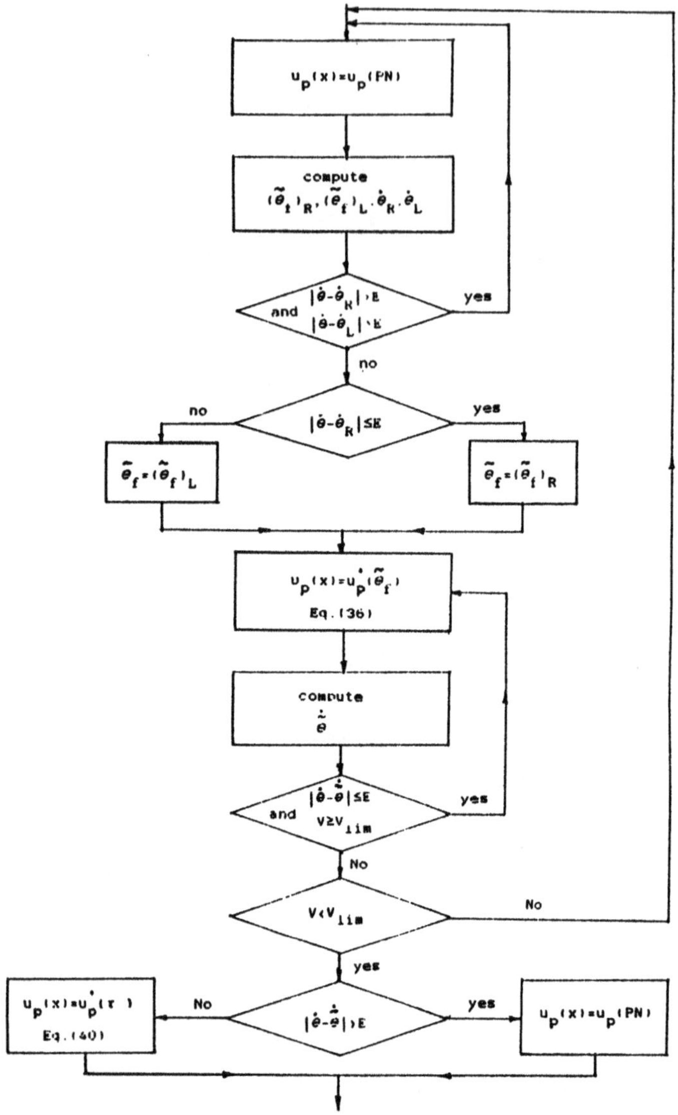

Fig. 5 Flowchart of feedback guidance law implementation.

REFERENCES

1. Guelman, M., Shinar, J., Green, A., "Qualitative Study of a Planar Pursuit Evasion Game in the Atmosphere", to be presented at the 1988 Guidance, Navigation, and Control Conference, to be held August 15-17 in Minneapolis, Minnesota.
2. Simakova, E.N., "The Quality Problem in Pursuit Games for Inertial Systems", Automation and Remote Control, Vol. 35, Nov. 1974, pp. 1716-1724.
3. Prasad, U.R., Rajan, N., Rao, N.J., "Planar Pursuit-Evasion with Variable Speeds, Part 1, Extremal Trajectory Maps", Journal of Optimization Theory and Applications, Vol. 33, No. 3, March 1981, pp. 401-418.
4. Guelman, M., "Proportional Navigation with a Maneuvering Target", IEEE Transactions on Aerospace and Electronic Systems, Vol. AES8, May 1977, pp. 364-371.
5. Shinar, J., "State Dependent Optimality of a Singular Subarc", Automatica, Vol. 24, March 1988.

PIECEWISE DETERMINISTIC DIFFERENTIAL GAMES*

A. Haurie**
GERAD, École des Hautes Études Commerciales, Montréal

Abstract: This paper deals with a class of stochastic differential games where the random disturbances are represented by stochastic jump processes. Between two successive random jump times the process is fully deterministic. It is possible to consider two classes of strategies, the usual feedback strategies and the so-called *piecewise open-loop* strategies in which an open-loop control is adapted by each player at a jump time. The characterization and the existence of equilibria are discussed for this class of games.

1. Introduction

The aim of this paper is to initiate the study a class of stochastic differential games where the random perturbations affecting the system are defined as a finite state jump process. Between two consecutive jumps of this process the system is deterministic with an evolution described by a state equation as in the usual definition of a deterministic differential game. As a motivating example one may imagine several economic agents sharing the same production system which is subject to random failure; the failure rates and the duration of repairs can depend on the total work processed on the system; between failures the system is deterministic. This description corresponds to a many-player version of the production planning model considered by Boukas and Haurie (1988).

Stochastic control systems with disturbances described as jump processes have been considered by Sworder (1969), Rishel (1975) and more recently by Vermes (1985) and Davis (1985). This last author coined the term *piecewise deterministic* systems. In the realm of differential games Basar and Haurie (1984), dealing with a class of feedback equilibria implying different leader-follower relationships, have extended the Rishel formalism, and the continuous time dynamic programming characterization of the solution to a many-player game setting. In the present paper one adopts a different approach for the definition of the so-called *piecewise deterministic differential games, (PDDG)*, closer in spirit to the work of Davis (1985) or to the modeling of production systems proposed by Boukas and Haurie (1988). One assumes that the players control a continuous-time system either through *piecewise open-loop* controls which can be adapted at the random jump times or through *feedback stationary strategies.* By this one means that at random jump times the state of the system is reset to some new initial value; this information is communicated to the players who have to define an open-loop control or a stationary feedback which will remain in force until the next random jump of the system occurs. In the one-player case, Davis (1985), Vermes (1985), Boukas and Haurie (1988) have exploited the relationship that exists between the optimal control of these systems and the solution of Markov decision processes in general (Borel) state and action spaces. In this paper one shows that, for the many-player case, it is also possible to establish a link between the two classes of piecewise deterministic differential games

* Aknowledgement: The author thanks Michèle Breton for her help in establishing the link with the sequential Markov game formalism

** Research supported by NSERC-Canada, Grant, A4952 and FCAR-Québec, Grants # 88EQ3528, and 89CE130

and noncooperative sequential Markov games in general state and action spaces, (recently studied by Whitt (1980), Breton (1987), Nowak (1987)), respectively.

The paper is organized as follows: In section 2 one considers the simple case of a *differential game with a random stopping time* and one gives a characterization of equilibria. In section 3 one introduces the more general concept of *PDDG*. In section 4 one recalls some known results concerning sequential Markov games with Borel state and action spaces. In section 5 one exploits the link which exists between this class of *PDDG*'s and the class of sequential Markov games in order to extend the results obtained in section 2. One also discusses existence results for noncooperative equilibria in *PDDG*'s. Finally, in section 6, one considers the formal extension of the Hamilton-Jacobi-Bellman equations to the characterization of feedback equilibria in *PDDG*'s.

2. A differential Game with a Random Stopping Time

In this section one considers the simple case of a deterministic system controlled by several players up to a random stopping time whose probability law depends on the system controls and trajectory. For this system one derives optimality conditions for open-loop equilibria in the form of a maximum principle, as well as the Hamilton-Jacobi-Bellman equations satisfied by a class of feedback equilibria.

2.1 Rules of the Game

Consider a dynamic system described by the state equation

$$\left.\begin{aligned} \dot{x}(t) &= f(x(t),\, \underline{u}(t)) \quad t \geq 0 \\ x(0) &= x^{\circ} \end{aligned}\right\} \tag{2.1}$$

where $M = \{1, \ldots, m\}$ is the set of players, $x \in \mathbb{R}^n$ is the state variable, $\underline{u} = (u_j)_{j \in M}$, and $u_j \in U_j \subset \mathbb{R}^{m_j}$ is the control variable of player j. One assumes that $f(x, \underline{u})$ is continuous and that there exists two numbers M and N such that for any x in $\mathbb{R}^n$

$$\|f(x, \underline{u})\| \leq M(1 + N\|x\|), \quad \forall \underline{u} \in \underline{U} = U_1 \times \cdots \times U_m.$$

Let $(\Omega, \mathcal{B}, P_{\underline{u}})$ be a probability space and let T be a **Markov Stopping Time**, s.t.

$$P_{\underline{u}}[T \in [t, t + dt) | T \geq t] = q(x(t), \underline{u}(t))\,dt + o(dt) \tag{2.2}$$

where $\lim\limits_{dt \to 0} \frac{o(dt)}{dt} = 0$ uniformely w.r.t. x and $\underline{u}$.

The function $q(x, \underline{u})$ is called the **intensity** of the stopping time; it is assumed to be as regular as $f(x, \underline{u})$, i.e. uniformly Lipshitz.

The payoff to player j is then defined by the functional

$$\begin{aligned} g_j(x^{\circ}, \underline{u}(\cdot)) = E_{\underline{u}(\cdot)} \Big[\int_0^T & e^{-\rho t} L_j(x(t), \underline{u}(t))\, dt \\ & + e^{-\rho T} \phi_j(x(T)) | x(0) = x^{\circ} \Big] \end{aligned} \tag{2.3}$$

where $L_j(x, \underline{u})$ is the reward rate functional of player j, and $\phi_j(x)$ is the terminal reward functional, assumed to be continuous and bounded. The discount rate ρ is positive, and $x(\cdot) : [0, \infty) \to \mathbb{R}^n$ is the unique trajectory emanating from x° and generated by the measurable control function $\underline{u}(\cdot) : [0, \infty) \to \underline{U}$, as the solution of (2.1).

The **rules of the game** are defined by (2.1)–(2.3). the game is played as follows: Each player j knows the rules; he observes the initial state x° and then picks an admissible control defined as a measurable function $u_j(\cdot) : [0, \infty) \to U_j$; the players strive to maximize their payoffs.

2.2 Reformulation as an Infinite Horizon Deterministic Differential Game.

In the game defined above the randomness is in the stopping time T only. If one gives an explicit formulation of the expected values determining the payoffs, and if one enlarges the state space it is possible to reformulate this game as an infinite horizon deterministic differential game. One has the following elementary probability

$$P\Big[T \in [t + dt)\Big] = q(x(t), \underline{u}(t))e^{-\int_0^t q(x(s), \underline{u}(s))ds}\, dt + o(dt), \tag{2.4}$$

therefore the payoffs (2.3) can be expressed as

$$g_j(x^{\circ}, \underline{u}(\cdot)) = \int_0^{\infty} \Big\{ \int_0^t \Big\{ e^{-\rho s} L_j(x(s), \underline{u}(s))\, ds + e^{-\rho t}\phi_j(x(t))\Big\} q(x(t), \underline{u}(t))e^{-\int_0^t q(x(s), \underline{u}(s))\, ds} \Big\}\, dt. \tag{2.5}$$

Due to the multiple integrals entering the expression (2.5) one does not still have a standard differential game format. This can be obtained if one introduces auxiliary state variables y_j, $\forall j \in M$, and z satisfying the following state equations

$$\dot{y}_j(t) = e^{-\rho t} L_j(x(t), \underline{u}(t)) \qquad y_j(0) = 0, \quad j \in M \tag{2.6}$$

$$\dot{z}(t) = q(x(t), \underline{u}(t)) \qquad z(0) = 0. \tag{2.7}$$

which permit one to redefine the payoffs as

$$G_j(X^{\circ}, \underline{u}(\cdot)) = \int_0^{\infty} \Big\{ y_j(t) + e^{-\rho t}\phi_j(x(t))\Big\} q(x(t), \underline{u}(t))e^{-z(t)}\, dt, \tag{2.8}$$

where $X^{\circ} = (x^{\circ}, 0, \ldots, 0)$ is the initial value for the extended state trajectory

$$X(\cdot) = (x(\cdot), y_1(\cdot), \ldots, y_m(\cdot), z(\cdot)) : [0, \infty) \mapsto \mathbb{R}^{n+m+1}.$$

Now the state equations defined by Eqs. (2.1), (2.6), (2.7) and the payoff functionals defined by Eq. (2.8) correspond exactly to the format of a deterministic differential game, played over an infinite time horizon.

2.3 An Equilibrium Maximum Principle

In this section one considers an **Open-Loop Equilibrium** (OLE) for the infinite horizon differential game defined above, and, under appropriate assumptions, one gives necessary conditions for optimality in the form of the **maximum principle** established by Halkin (1974) for infinite horizon optimal control problems. Then, under additional assumptions concerning the asymptotic behavior of adjoint variables one is able to express this maximum principle in terms of the original variables entering the definition of the rules of the game given in section 2.1.

Definition 2.1. *A control vector function $\underline{u}^*(\cdot) = (u_1^*(\cdot), \ldots, u_m^*(\cdot))$ is an Open-Loop Equilibrium (OLE) at x^o for the game defined in section 2.1 if, for any player $j \in M$, and any admissible control function $u_j(\cdot) : [0, \infty) \to U_j$ the following holds*

(i) $u_j^*(\cdot) : [0, \infty) \to U_j, \quad \forall j \in M$ (*i.e. the equilibrium is admissible*)

(ii) $\forall j \in M, \quad \forall u_j(\cdot) : [0,\infty) \mapsto U_j \quad g_j(x_o, \underline{u}^{*(j)}(\cdot)) \le g_j(x^o, \underline{u}^*(\cdot)),$ (2.9)
where one adopts the notation
$\underline{u}^{*(j)}(\cdot) = (u_1^*(\cdot), \ldots, u_{j-1}^*(\cdot), u_j(\cdot), u_{j+1}^*(\cdot), \ldots, u_m^*(\cdot)).$

Remark 2.1. *Even if one is dealing with a stochastic game, the use of open-loop strategies makes sense since, up to the random stopping time T, the system is totally deterministic. As the sample set of the stopping time T is $[0, \infty)$, the open-loop controls have to be defined over an infinite time horizon.*

Remark 2.2. *An OLE $\underline{u}^*(\cdot)$ is also an OLE for the associated infinite horizon deterministic differerential game defined in Section 2.3.*

In order to invoke Halkin's (1974) Maximum Principle one has to assume that f, L and q are C^1 in x with bounded derivatives. Consider the hamiltonians

$$\begin{aligned} H_j(\underline{u}, X, \mu_o^j, \underline{\mu}^j, t) = {} & \mu_o^j(y_j + e^{-\rho t}\phi_j(x))q(x, \underline{u})e^{-z} \\ & + \mu_x^j f(x, \underline{u}) + \mu_{y_j}^j e^{-\rho t} L_j(x, \underline{u}) + \mu_z^j q(x, \underline{u}), \ j \in M, \end{aligned} \tag{2.10}$$

where

$$\mu_o^j \in \mathbb{R}, \quad \underline{\mu}^j = (\mu_x^j, \mu_{y_j}^j, \mu_z) \in \mathbb{R}^{n+2}$$

Proposition 2.1. *If $\underline{u}^*(\cdot) = (u_j^*(\cdot))_{j \in M}$ is an equilibrium at x^o, generating a trajectory $X^*(\cdot) = (x^*(\cdot), y^*(\cdot), z^*(\cdot))$ emanating from $X^o = (x^o, 0, 0)$, then there exist a scalar $\mu_o^j \ge 0$ and m absolutely continuous functions $\underline{\mu}^j(\cdot) : [0, \infty) \to \mathbb{R}^{n+2}$ such that, $\forall t \ge 0, \quad \forall j \in M$*

(i)

$$H_j(\underline{u}^*(t), X^*(t), \mu_o^j, \underline{\mu}^j(t), t) = \sup_{u_j \in U_j} H_j\left([\underline{u}^{*(-j)}(t), u_j], X^*(t), \mu_o^j, \underline{\mu}^j(t), t\right) \tag{2.11}$$

where

$$[\underline{u}^{*(-j)}(t), u_j] = (u_1^*(t), \ldots, u_{j-1}^*(t), u_j, u_{j+1}^*(t), \ldots, u_m^*(t)),$$

(ii)

$$\dot{\underline{\mu}}^j(t) = -\nabla_X H_j(\underline{u}^*(t), X^*(t), \mu_o^j, \underline{\mu}^j(t), t) \quad \text{a.e.} \quad \text{for } t \ge 0, \tag{2.12}$$

(iii)

$$(\mu_o^j, \underline{\mu}^j(0)) \ne 0. \tag{2.13}$$

Proof: See Halkin (1974). ∎

Halkin has also shown that it is impossible to obtain as general necessary conditions that $\mu_0^j > 0$, and that $\lim_{t \to \infty} \underline{\mu}^j(t) = 0$. If these conditions are also satisfied one can reformulate the maximum principle, in terms of the original state variable x only. For that purpose one introduces the **modified hamiltonians**, for $j \in M$

$$\mathcal{H}_j(x, \underline{u}, \lambda^j, \mathcal{L}_j) = L_j(x, u) + \lambda^j f(x, u) + (\phi_j(x) - \mathcal{L}_j)q(x, u) \tag{2.14}$$

and, along the equilibrium trajectory $x^*(\cdot)$ one defines $\mathcal{L}_j^*(t)$ as the expected payoff from time t onward, discounted at t, for player j, given that no jump as occured up to time t. More precisely

$$\mathcal{L}_j^*(t) = \int_0^\infty \Big\{ \int_0^\tau e^{-\rho s} L_j(x^*(t+s), \underline{u}^*(t+s))\, ds + e^{-\rho t} \phi_j\big(x^*(t+\tau)\big)\Big\} q(x^*(t+\tau), \underline{u}^*(t+\tau))\, e^{-\int_0^\tau q\left(x^*(t+s), \underline{u}^*(t+s)\right) ds}\, d\tau. \tag{2.15}$$

Proposition 2.2. *If, for each player $j \in M$, $\mu_0^j \neq 0$ and $\underline{\mu}^j(t) \to 0$, as $t \to \infty$, then there exist m absolutely continuous functions $\lambda^j : [0, \infty) \to \mathbb{R}^n$, such that the following holds:*

(i)

$$\mathcal{H}_j(x^*(t), \underline{u}^*(t), \lambda^j(t), \mathcal{L}_j^*(t)) = \sup_{u_j \in U_j} \mathcal{H}_j(x^*(t), [\underline{u}^{*(-j)}(t), u_j], \lambda^j(t), \mathcal{L}_j^*(t)) \tag{2.16}$$

(ii)

$$\dot{\lambda}^j(t) = -\nabla_x \mathcal{H}_j(x^*(t), \underline{u}^*(t), \lambda^j(t), \mathcal{L}_j^*(t)) + \big(\rho + q(x^*(t), u^*(t))\big)\lambda^j(t). \tag{2.17}$$

Proof. If one explicits the adjoint variational equations (2.11) one obtains

$$\frac{d}{dt}\mu_{y_j}^j(t) = \frac{d}{dt}[e^{-z^*(t)}] \tag{2.18}$$

$$\frac{d}{dt}\mu_z^j(t) = \big(y_j^*(t) + e^{-\rho t}\phi_j(x^*(t))\big) q(x^*(t), \underline{u}^*(t)) e^{-z^*(t)} \tag{2.19}$$

and thus, by direct integration

$$\mu_{y_j}^j(t) = e^{-z^*(t)} \tag{2.20}$$

$$\mu_z^j(t) = e^{-z^*(t)}\,(y_j^*(t) + e^{-\rho t}\mathcal{L}_j^*(t)) \tag{2.21}$$

with $\mathcal{L}_j^*(t)$ defined as in (2.15). Now, if one defines

$$\lambda^j(t) = e^{z^*(t)+\rho t}\,\mu_x(t) \tag{2.22}$$

and if one substitutes into (2.11)–(2.13) one readilly obtains the conditions (2.14)–(2.17). ∎

Remark 2.3. *The modified hamiltonians $\mathcal{H}_i$ are reminiscent of the so-called current-value hamiltonians used to express the maximum principle of stationary systems with discounting.*

Remark 2.4. *$\mathcal{L}_j^*(t)$ could also be called the current-value "reward-to-go" function along the equilibrium state trajectory. With this in mind the modified hamiltonian can be interpreted in terms of trade-offs between immediate reward, $L_j(x, \underline{u})$, state improvement, $\lambda^j f(x, \underline{u})$, and net reward due to a jump, $(\phi_j(x) - \mathcal{L})q(x, \underline{u})$.*

Remark 2.5. *For a discussion of general conditions insuring that the required asymptotic transversality conditions hold, the reader is referred to Boukas et al. (1988).*

2.4 Feedback Equilibrium and Dynamic Programming Equations.

The game can also be played with a feedback information structure, meaning that each player has access to the state value x, at each instant of time. Therefore he can control the system through a stationary policy $\gamma_j : \mathbb{R}^n \to U_j$, according to which he will choose the action $u_j = \gamma_j(x)$ if he observes that the current state is x. Given a policy vector $\underline{\gamma} = (\gamma_j)_{j\in M}$, the feedback controlled system is described by the equations

$$\left.\begin{aligned} \dot{x}(t) &= f\big(x(t), \underline{\gamma}(x(t))\big) \quad t \geq 0 \\ x(0) &= x^o \end{aligned}\right\} \tag{2.23}$$

$$P[T \in [t, t+dt) | T \geq t] = q\big(x, \underline{\gamma}(x(t))\big)dt + o(dt). \tag{2.24}$$

$$V_j(x^o, \underline{\gamma}) = E_{\underline{\gamma}}\Big[\int_0^T e^{-\rho t} L_j\big(x(t), \underline{\gamma}(x(t))\big)dt + e^{-\rho T}\phi_j(x(T)) | x(0) = x^o\Big] \tag{2.25}$$

Of course, to be meaningful, these expressions require $\underline{\gamma}$ to be sufficiently regular. One refers the reader to Flemming and Rishel (1979), Boltyanskii (1966), Mirica (1969), Rishel (1975b) for a discussion of these conditions. Another possibility would be to use the δ-game approach proposed by Friedman (1971) in order to define in a mathematically correct sense a differential game with this imformation structure. For the ease of exposition one implicitly assumes that enough regularily assumptions hold.

Definition 2.2. *A policy vector $\underline{\gamma}^* = (\gamma_j^*)_{j\in M}$ is a stationary feedback equilibirum of the game defined in Section 2.1 if for any $j \in M$ and any admissible vector*

$$\underline{\gamma}^{*(j)} = (\gamma_1^*, \ldots, \gamma_{j-1}^*, \gamma_j, \gamma_{j+1}^*, \ldots, \gamma_m^*)$$

and for any $x \in \mathbb{R}^n$ one has

$$V_j^*(x) = V_j(x, \underline{\gamma}^*) \geq V_j(x, \underline{\gamma}^{*(j)}). \tag{2.26}$$

Under the assumption of continuity and piecewise differentiability of the **value functionals** *$V_j^*(x)$ one can characterize a feedback equilibirum through a set of* **Hamilton-Jacobi-Bellman** *(HJB) equations. Omitting the details of derivation which would be very similar to those in Basar and Haurie (1974) one can state the result*

Proposition 2.3. *If $\underline{\gamma}^* = (\gamma_j^*)_{j\in M}$ is a feedback equilibirum which satisfies the regularity conditions of Boltyanskii or Mirica, then the value functionals $V_j^*(x)$ defined in (2.26) satisfy the following set of HJB equations*

$$\rho V_j^*(x) = \sup_{u_j\in U_j} \Big\{L_j(x, [\underline{\gamma}^{*(-j)}(x), u_j]) + \frac{\partial}{\partial x}V_j^*(x)\, f(x, [\underline{\gamma}^{*(-j)}(x), u_j]) \\ + (\phi_j(x) - V_j^*(x))q(x, [\underline{\gamma}^{*(-j)}(x), u_j])\Big\} \quad j \in M. \tag{2.27}$$

where

$$[\underline{\gamma}^{*(-j)}(x), u_j] = (\gamma_1^*(x), \dots, \gamma_{j-1}^*(x), u_j, \gamma_{j+1}^*(x), \dots, \gamma_m^*(x)).$$

Remark 2.5. *As in the case of purely deterministic differential games the feedback equilibria do not coincide in general with the OLE's. One should however note the similitude between the modified hamiltonians (2.14) and the right-hand-side of the HJB equations. Recall also the interpretation of $\mathcal{L}^*(t)$ as the "reward-to-go" along the equilibrium trajectory.*

3. Stochastic differential game with jump Markov disturbances

In this section one describes a more general class of *PDDG*'s, consisting of stochastic games played on a dynamical system controlled by m players, and subject to random perturbations defined as jump Markov processes.

At any time $t \geq 0$ the system state is given by $s(t) = (x(t), y(t))$, where $x(t) \in \mathbb{R}^n$, and $y(t) \in E$, where E is a finite set. With each $k \in E$ there are associated m nonempty control sets $U_j^k \subset \mathbb{R}^{p_j^k}$, $j \in M$, p_j^k positive integer, and a *state equation*

$$\dot{x}(t) = f^k(x(t), u_1(t), \dots, u_m(t)) \tag{3.1}$$

where the function $f^k : \mathbb{R}^n \times U_1^k \times \cdots \times U_m^k \mapsto \mathbb{R}^n$ is continuous and satisfies the following assumptions

(A1) There exist two numbers A^k and B^k such that $\forall x \in \mathbb{R}^n$, and $\forall \underline{u} = (u_1, \dots, u_m) \in U_1^k \times \cdots \times U_m^k$ the following holds

$$f^k(x, \underline{u}) \leq A^k(1 + B^k\|x\|).$$

This assumption insures that, given an initial state x^k and m controls defined as measurable functions $u_j(\cdot) : [0, \infty) \mapsto U_j^k$, $j \in M$, there exists a unique trajectory $x(\cdot) : [0\infty) \mapsto \mathbb{R}^n$ such that

$$x(0) = x^k \\ \dot{x}(t) = f^k(x(t), \underline{u}(t)) \quad \text{a.e. on } [0, \infty). \tag{3.2}$$

Let $(\Omega, \mathcal{B})$ be a measurable space and $T : \Omega \mapsto [0, \infty)$ a random variable, called the *time to go until the next jump*. One assumes that

(A2) There exist bounded jump rates. i.e.

$$P[T \in [t, t + dt)|T \geq t;\, y(t), x(t), \underline{u}(t)] = q_{y(t)}(x(t), \underline{u})\, dt + o(dt) \tag{3.3}$$

where, for each $k \in E$, $q_k(x,\underline{u})$ is a strictly positive continuous function on $\mathbb{R}^n \times U_1 \times \cdots \times U_m$ and $\frac{o(dt)}{dt} \to 0$ with dt uniformly in $x, \underline{u}$.

Finally let $\Phi(\cdot,\cdot|x(t),y(t))$ be a conditional probability measure on $\mathbb{R}^n \times E$, given that a jump has occured at time t, and that the system was in state $(x(t),y(t))$. This conditional probability measure is called the *random reset function* and its role will be described shortly. One assumes that

(A3) the random reset function is continuous w.r.t. $x, y, \underline{u}$.

The game is played as follows. At initial time $t^0 = 0$ the state of the system $s^0 = (x^0, y^0)$ is given and observed by the m players. Each player selects a measurable control $u_j(\cdot) : [0,\infty) \mapsto U_j^{y^0}$ in the class $\mathcal{U}_j^{y^0}$ of such functions. It then corresponds a unique solution $x(\cdot) : [0,\infty) \mapsto \mathbb{R}^n$ of the differential equation (3.1) with initial condition $x(0) = x^0$. If at a random time τ^1 the process jumps, then the jump time is recorded, $t^1 = t^0 + \tau^1 = \tau^1$, and the state is reset to a new value (x^1, y^1) picked randomly in $\mathbb{R}^n \times E$ according to the probability measure $\Phi(\cdot,\cdot|y^0, x(t^1))$. Then each player observes (x^1, y^1) and selects a new measurable control $u_j : [0,\infty) \mapsto U_j^{y^1}$ in the class $\mathcal{U}_j^{y^1}$. This control vector, with the initial condition $x(0) = x^1$, generates a unique solution $x(\cdot) : [0,\infty) \mapsto \mathbb{R}^n$ of the differential equation (3.1), etc... The successive random jump times will thus be defined as $t^\nu = \sum_1^\nu \tau^\nu$, and the successive random states observed at jump times are defined as $s^\nu = (x^\nu, y^\nu)$, $\nu = 0, 1, \ldots$

For each player $j \in M$ and each $k \in E$, let $L_j^k : \mathbb{R}^n \times U_1^k \times \cdots \times U_m^k \mapsto \mathbb{R}$ and $G_j^k : \mathbb{R}^n \mapsto \mathbb{R}$ be two bounded and continuous functions. If, at jump time t^ν, the system is in state $s^\nu = (x^\nu, y^\nu)$, if the m players choose the controls $u_j(\cdot)$, $j \in M$, which generate the state trajectory $x(\cdot)$, then the **transition reward** to player j between the two consecutive jump times t^ν and $t^{\nu+1} = t^\nu + \tau^{\nu+1}$ is given by

$$g_j(s^\nu, \underline{u}(\cdot)) =$$
$$E_{\underline{u}(\cdot)}\Big[\int_0^{\tau^{\nu+1}} e^{-\rho t} L_j^{y^\nu}(x(t), \underline{u}(t))\,dt + G_j^{y^\nu}(x(\tau^{\nu+1}))|x(0) = x^\nu\Big] \tag{3.4}$$

where ρ is a positive discount rate.

One assumes that the players want to maximize the sum of their expected discounted transition rewards over an infinite time horizon.

4. Equilibria in a sequential Markov game

In this section one recalls some results concerning the noncooperative equilibria in the class of sequential Markov games with Borel state and action spaces.

A m-player sequential Markov game is defined by the four-tuple (S, Γ, r_M, Q) where

- S is a *Borel* state space;
- $\Gamma(\cdot)$ is a Borel measurable mapping that associates with each $s \in S$ the product control set $A_1(s) \times \cdots \times A_m(s)$, where $A_j(s)$ is the non empty *action set* of Player j, admissible at state s;
- $r_M = (r_j(\cdot,\cdot))_{j\in M}$ is the reward function vector where $r_j(s, a_1, \ldots, a_m)$ is the expected transition reward for player j when the system is in state s and the players use the actions $a_1, \ldots, a_m$ respectively;

- Q is the transition probability kernel ; given the state s^ν at jump time t^ν, and given the actions $a_1, \ldots, a_m$, $Q(ds, d\tau | s^\nu, a_1, \ldots, a_m)$ is the elementary probability that the next jump occur in the time interval $[\tau, \tau + d\tau)$ and that the new state be $s' \in ds$; for any Borel subset S' of $S \times [0, \infty)$, $Q(S'|s, a_1, \ldots, a_m)$ is supposed to be a measurable function of $(s, a_1, \ldots, a_m)$.

A *policy* δ_j is a Borel measurable mapping from S into the action set of player j, such that, whenever state $s \in S$ is observed , player j chooses the action $a_j = \delta_j(s) \in A_j(s)$. One denote by Δ_j the set of all policies for player j, and by $\Delta = \prod_{j \in M} \Delta_j$ the product set of policy vectors. To a policy corresponds a *stationary strategy* which dictates player j's action choices for all periods in the sequential game. The Ionescu-Tulsea theorem assures that, under the measurability assumptions made above, associated with an initial state $s \in S$ and a stationary strategy vector $\delta \in \Delta$, there exists a uniquely defined probability measure P_{δ, s^0} over the set of sequences $(t^0, s^0, a^0, t^1, s^1, a^1, \ldots, t^\nu, s^\nu, a^\nu, \ldots)$ of jump times, states and actions. The *value function* associated with δ and s^0 is then defined as

$$v_\delta(s^0, j) = E_{\delta, s}[\sum_{\nu=0}^{\infty} e^{-\rho t^\nu} r_j(s^\nu, a_1^\nu, \ldots, a_m^\nu)]. \tag{3.5}$$

Let $\mathcal{V}$ be the Banach space of bounded continuous functions $v(\cdot) : S \times M \mapsto \mathbb{R}$ endowed with the sup norm. As in Denardo (1967) one defines the local return function

$$h(s, j, a_1, \ldots, a_m, v(\cdot)) = r_j(s, a_1, \ldots, a_m) + \int_{S \times [0,\infty)} e^{-\rho \tau} v(s', j) Q(ds', d\tau | s, a_1, \ldots, a_m),$$

and for a given policy vector $\delta \in \Delta$, one introduces the operator $H_\delta : \mathcal{V} \mapsto \mathcal{V}$ defined by

$$(H_\delta v(\cdot))(s, j) = h(s, j, \delta(s), v(\cdot)). \tag{3.6}$$

As in Whitt (1980) one also considers the *optimal response* operator $F_\delta : \mathcal{V} \mapsto \mathcal{V}$ defined by

$$(F_\delta v(\cdot))(s, j) = \sup_{a_j \in A_j(s)} h(s, j, [\delta^{-j}(s), a_j], v(\cdot)), \tag{3.7}$$

where $[\delta^{-j}(s), a_j]$ is a vector of actions a' such that $a'_i = \delta_i(s)$ if $i \neq j$ and $a'_j = a_j$.

It is convenient to define the *sub-probability kernel*

$$\mathcal{Q}(ds'|s, a_1, \ldots, a_m) = \int_{[0,\infty)} e^{-\rho \tau} Q(ds', d\tau | s, a_1, \ldots, a_m). \tag{3.8}$$

It is then easy to check that if the reward function r_M is bounded and if $\|\mathcal{Q}\| \leq c < 1$ then the operators H_δ and F_δ are bounded, monotone and contracting and thus admit unique fixed points denoted $v_\delta(\cdot)$ and $f_\delta(\cdot)$ respectively. $v_\delta(\cdot, j)$ is the value function for player j associated with the stationary strategy vector δ, while $f_\delta(\cdot, j)$ is the *optimal* value function for player j when the other players use the stationary strategy vector δ, i.e.

$$f_\delta(s, j) = \sup\{v_{[\delta^{-j}, \gamma_j]}(s, j);\ \gamma_j \in \Delta_j\}$$

where $[\delta^{-j}, \gamma_j]$ is a stationary strategy vector δ' such that $\delta'_i = \delta_i$ if $i \neq j$ and $\delta'_j = \gamma_j$.

Definition 4.1. *A stationary strategy is an ε-equilibrium ($\varepsilon \geq 0$) if*

$$f_\delta(s, j) - v_\delta(s, j) \leq \varepsilon \ \forall s \in S, \ j \in M. \tag{3.9}$$

A 0-equilibrium is simply called an equilibrium.

If one introduces the point to set mapping $\Psi_\varepsilon : \Delta \mapsto 2^\Delta$ defined by

$$\Psi_\varepsilon(\delta)_j = \{\gamma_j : f_\delta(s,j) \leq v_{[\delta-j,\gamma_j]}(s,j) + \varepsilon \,;\, \forall s \in S\} \,, \forall j \in M, \tag{3.10}$$

then a ε-equilibrium defined by a stationary strategy δ^* is also a *fixed point* of Ψ_ε, i.e. $\delta^* \in \Psi_\varepsilon(\delta^*)$.

The existence of an *equilibrium* ($\varepsilon = 0$) can be established through the following theorem (one uses the notation Ψ instead of Ψ_0);

Theorem 4.1 (Kakutani, Gliksberg, Fan). *Assume*

(1) Δ is a convex compact subset of a Hausdorff locally convex topological vector space;

(2) $\Psi : \Delta \mapsto 2^\Delta$ is convex valued;

(3) Ψ is upper semicontinuous.

Then there exists an equilibrium given by a stationary strategy vector δ^*.

Whitt (1980) has shown that this theorem admits the following corollary

Theorem 4.2. *Assume*

(1) $A_j(s)$ is a convex compact subset of a locally compact topological vector space (LCTVS);

(2) $h(s,j,a,v(\cdot))$ is a concave function of a_j for each s, j, $v(\cdot)$, and

(3) $v_\delta(s,j)$ and $f_\delta(s,j)$ are continuous functions of δ for each $s \in S$ and $j \in M$.

Then there exists an equilibrium given by a stationary strategy vector δ^*.

5. Existence and Characterization of ε-Equilibria in PDDG's

In this section one establishes a formal link between the class of *PDDG*'s and the sequential Markov games. Invoking the theorems stated in section 4 on then gives a set of conditions under which the *PDDG*'s admit ε-equilibria. These equilibria are formulated in the class of mixed stationary strategies. If the equilibrium exists in the class of pure stationary strategies then one can extend the optimality conditions obtained un section 2.

5.1. PDDG in a Sequential Markov Game Format

Define the four-tuple (S, Γ, r_M, Q) where

- $S = E \times \mathbb{R}^n$;
- $\Gamma(\cdot)$ is the mapping that associates with each $s \in S$ the product control set $A_1(s) \times \cdots \times A_m(s)$, where $A_j(s)$ is the class of probability measures $\mathcal{P}[\mathcal{U}_j^y]$ on the set $\mathcal{U}_j^y$ of admissible measurable controls $u_j : [0,\infty) \mapsto U_j^y$;

- $r_M = (r_j(\cdot,\cdot))_{j\in M}$ is the reward function vector where

$$r_j(s,a_1,\ldots,a_m) = E_{(a_1,\ldots,a_m)}[g_j(s,\,\underline{u}(\cdot))\,];$$

where $g_j(s,\,\underline{u}(\cdot))$ is defined as in (3.4)

- $$Q(ds|s,a_1,\ldots,a_m) = \int_{\prod_{j\in M}U_j^y} d\mu_{a_1}(u_1(\cdot)),\ldots,d\mu_{a_m}(u_m(\cdot))\int_0^\infty e^{-\rho t}\Phi(ds|x(t),y)$$
$$q_y(x(t),u_1(t),\ldots,u_m(t))e^{-\int_0^t q_y(x(s),u_1(s),\ldots,u_m(s))\,ds}\,dt$$

These definitions permit one to recast the piecewise deterministic differential game, introduced in section 2, in the format of a sequential game. The state set S and the action sets $A(s)$ are infinite dimensional spaces. As usual in game theory one considers the class of so-called *mixed startegies.* This implies that the players select randomly the control function they will use between two consecutive jump times t^ν and $t^{\nu+1}$. This randomization process plays a fundamental role in the proof of existence of equilibria for *PDDG*'s. As in Whitt (1980) one first gives conditions for existence of equilibria in the case of a countable state set; then one approximates the continuous state space game through a sequence of countable state games and this permits one to establish existence of ε-equilibria in general PDDG's.

5.2. Existence of equilibria for a PDDG with a countable reset function range.

In addition to A1-A3 assume the following

(A4) the probability measure $\Phi(ds|x(t),y)$ is concentrated on a countable subset $D = E\times X$ of S.

Theorem 5.1. *If assumptions A(1)-A(4) hold then there exists an equilibrium in mixed strategies for the PDDG.*

Proof: The proof consists in checking that the three conditions of Theorem 4.2 hold.

(1) For each $k\in E$ and $j\in M$ one can restrict the class of admissible controls to

$$L^2_{loc}[[0,\infty);U_j^k];$$

therefore the set $\mathcal{U}_j^k$ is a separable compact metric space, and the set $A_j(s)$ of *mixed admissible actions* for player j at state s is also a convex compact subset of a *LCTVS.*

(2) The concavity of $h(s,j,a,v(\cdot))$ is a direct consequence of the randomization process;

(3) Since the expression

$$E_{\underline{u}(\cdot)}[\int_0^\tau e^{-\rho t}L_j^y(x(t),u_1(t),\ldots,u_m(t))\,dt + e^{-\rho\tau}G_j^y(x(\tau)|s(0)=s]$$

is continuous w.r.t. $u_j(\cdot)$, $j\in M$, as well as the expression

$$\int_0^\infty e^{-\rho t}\Phi(ds|x(t),y)q_y(x(t),u_1(t),\ldots,u_m(t))e^{-\int_0^t q_y(x(s),u_1(s),\ldots,u_m(s))\,ds}\,dt,$$

it can be deduced that $r_j(s,a)$ and $Q(A|s,u(\cdot))$ are continuous in a for each $j \in M$ and $s \in S$. Since the probability measures Φ are concentrated on a countable subset D of S, from the point of view of the sequential discrete stage game, the state set can be reduced to D. In that case Whitt (1980), Theorem 3, shows that $v_\delta(s,j)$ and $f_\delta(s,j)$ are continuous in δ for each (s,j). Therefore all conditions of Theorem 4.2. are satisfied and there exists an equilibrium.

5.3. Existence of ε-Equilibria in a *PDDG*

One has to assume the following

(A4′) the state x remains in a bounded subset X of $\mathbb{R}^n$.

Theorem 5.2. *If assumptions A(1)-A(4′) hold, for any $\varepsilon \geq 0$ there exists an ε-equilibrium in mixed strategies for the PDDG.*

Proof: One follows closely the approach proposed by Whitt (1980). Let $(X_1,\dots,X_\ell,\dots)$ be a countable partition of X and let x_ℓ be a selected point in each subset X_ℓ. Let $\tilde{X} = \{x_1,\dots,x_\ell,\dots\}$ be the set of these representatives and let $\tilde{S} = \tilde{X} \times E$ be the *restricted state set.* One then defines an approximation of the *PDDG* by the data $(\tilde{S},\Gamma,r_M,\tilde{Q})$, where, for any $s \in \tilde{S}$ and $s' = (x_\ell, y) \in \tilde{S}$ one defines

$$\tilde{Q}(s'|s,a) = Q(X_\ell, y|s,a).$$

This approximate game corresponds to a *PDDG* with a countable range for the reset function. According to Theorem 5.1. there exists an equilibrium defined by a stationary mixed strategy vector $\tilde{\delta}^*$, with an associated value function $\tilde{v}_{\tilde{\delta}^*}$. Let $e(\tilde{v}_{\tilde{\delta}^*})$ be the extrapolation of $\tilde{v}_{\tilde{\delta}^*}$ to a piecewise constant function on $X \times E$:

$$e(\tilde{v}_{\tilde{\delta}^*})(x,y;j) = \tilde{v}_{\tilde{\delta}^*}(x_\ell,y;j) \quad \forall x \in X_\ell.$$

Then, by an approximation result obtained by Whitt (1978), it can be shown that $\tilde{\delta}^*$ is a $\frac{1}{1-\rho}2K(\tilde{v}_{\tilde{\delta}^*})$-equilibrium, with

$$K(\tilde{v}_{\tilde{\delta}^*}) = \sup_{j\in M, s\in S} \{h(s,j,\tilde{\delta}^*(s),e(\tilde{v}_{\tilde{\delta}^*})) - \tilde{h}(p(s),j,\tilde{\delta}^*(s),\tilde{v}_{\tilde{\delta}^*})\},$$

with $p(s) = (x_\ell, y)$ if $s = (x,y)$, $x \in X_\ell$. As the diameter of the partition tends to 0 $K(\tilde{v}_{\tilde{\delta}^*})$ tends also to 0, hence the result. ∎

5.4. Dynamic Programming Equations for Pure POL Equilibria

A pure strategy equilibrium in a *PDDG* involves piecewise open-loop (POL) controls. Such an equilibrium is characterized by a dynamic programming equation which is related to the maximum principle derived in section 2 for the class of differential games with a random stopping time (*DGRST*).

If an equilibrium exists in the class of pure strategies then for each player the strategy δ_j^* associates a control $u_j(\cdot) \mapsto U_j^{y^\nu}$ with an observed state (x^ν, y^ν) at a jump time t_ν. Since at an equilibrium one must have

$$v_{\delta^*} = f_{\delta^*} = v^*$$

one obtains from (3.7) that

$$v^*(s,j) = \sup_{u_j(\cdot)} h(s,j,[\delta^{*-(j)}(s),u_j(\cdot)],v^*(\cdot)), \quad \forall j \in M\,,\, s \in S. \tag{5.1}$$

For any $s \in S$, the right-hand-side of Eq.(5.1) corresponds to an OLE for a *DGRST*, if one defines the *terminal reward* functions

$$\phi_{j,s}(x(\tau)) = G_j^y(x(\tau)) + \int_S \Phi(ds'|x(\tau,y)v^*(s',j). \tag{5.2}$$

Therefore the dynamic programming equations (3.8)-(3.10) are expressed in terms of a class of local games, defined for each possible state s at a jump time, and each of this local game is a *DGRST*.

6. Equilibria in the class of feedback strategies

It would be straightforward to obtain the HJB equations characterizing a *feedback equilibrium* in a *PDDG*. A feedback strategy for player j is a mapping $\underline{\gamma}_j : (x,y) \mapsto U_j^y$. Introducing, for each player $j \in M$, the Bellman value function $W_j^*(t,x,y) = e^{-\rho t}V_j^*(x,y)$, one would obtain the following set of HJB equations characterizing an equilibrium feedback strategy vector $\underline{\gamma}^*$

$$\begin{aligned}\rho V_j^*(x,y) = \sup_{u_j \in U_j^y} \Big\{ & L_j(x,\, [\underline{\gamma}^{*(-j)}(x,y),\, u_j]) + \frac{\partial}{\partial x} V_j^*(x)\, f(x,\, [\underline{\gamma}^{*(-j)}(x,y),\, u_j]) \\ & + [G_j^y(x) + \int_S \Phi(ds'|x(\tau,y)\{V_j^*(x',y') - V_j^*(x,y)\}]q(x,\, [\underline{\gamma}^{*(-j)}(x,y),\, u_j])\Big\} \quad j \in M.\end{aligned}$$

where

$$[\underline{\gamma}^{*(-j)}(x,y),\, u_j] = (\gamma_1^*(x,y), \ldots, \gamma_{j-1}^*(x,y), u_j, \gamma_{j+1}^*(x,y), \ldots, \gamma_m^*(x,y)).$$

7. Conclusion

The consideration of *PDDG*'s opens several avenues of research. This class of games provides an interesting paradigm which should prove useful in the modeling of economic competition in the presence of possible random switches in the mode of play. On a pure game theoretic basis this class of games offers a nontrivial example of generalized Markov sequential game. The concept of randomized strategy which appears naturally in this class of games has yet to be given a practical significance through the establishment of an approximate computational technique.

References

T. Basar and A. Haurie, 1984, "Feedback Equilibra in Differential Games with Structural and Modal Uncertainties", *Advances in large scale systems*, Editor: Jose B. Gruz Jr., Vol. 1, pp. 163–201.

V.G. Boltyanskii, 1966, "Sufficient Conditions For Optimality And The Justification Of The Dynamic Programming Method", *SIAM J. Control*, Vol. 4, no. 2, pp. 326–361.

E.K. Boukas and A. Haurie, 1987, "Optimality Conditions For Continuous Time Systems With Controlled Jump Markov Disturbances: Application To Production And Maintenance Scheduling ", in, A. Bensoussan, J.L. Lions edit., Proceeding INRIA 8th International Conference Springer Verlag on *Analysis and Optimization of Systems*, Antibes June 1988.

E.K. Boukas, A. Haurie and P. Michel, 1988, "An Optimal Control Problem with a Random Stopping Time", *Cahier du GERAD.*

M. Breton, 1987, "Équilibres pour des jeux séquentiels", PhD. thesis, Université de Montréal.

M. H. A. Davis, 1985, Control of Piecewise-Deterministic Processes Via Discrete-Time Dynamic Programming, in Proceedings of 3th Bad Honnef Symposium on Stochastic Differential Systems, 1985, Lectures Notes on Control and Inf-Sciences vol. 78.

E.V. Denardo, 1967, "Contraction Mappings in the Theory Underlying Dynamic Programming", *SIAM Review*, Vol. 9, no. 2, pp. 165–177.

A. Friedman, 1971, *Differential Games.* Wiley-Interscience, New York

H. Halkin, 1974, "Necessary Conditions For Optimal Control Problems With Infinite Horizons", *Econometrica*, Vol. 42, no. 2, pp. 267–272.

S. Mirica, 1969, "On The Admissible Synthesis In Optimal Control Theory And Differential Games", *SIAM J. Control*, Vol. 7, no. 2, pp. 292–316.

A.S. Nowak, "Nonrandomized Strategy Equilibria in Noncooperative Stochastic Games with Additive Transition and Reward Structure", *Journal of Optimization Theory and Applications*, Vol. 52, no. 3, pp 429-441.

R. Rishel, 1975 a), "Dynamic Programming And Minimum Principles for Systems With Jump Markov Disturbances", *SIAM J. Control*, Vol 13, no. 2, pp.338–371.

R. Rishel, 1975 b), "Control Of Systems With Jump Markov Disturbances", *IEEE, Trans. on Automatic Control*, Vol. AC-20, no. 2,

D. Vermes, 1985, "Optimal Control of Piecewise Deterministic Markov Proces", *Stochastics*, 14, pp. 165–208.

W. Whitt, 1978, "Approximation of Dynamic Programs I", *Math. Oper. Res.* Vol. 3, pp. 231–243.

W. Whitt, 1979, "Approximation of Dynamic Programs II", *Math. Oper. Res.* Vol. 4, pp. 179–185.

W. Whitt, 1980, "Representation and Approximation of Noncooperative Sequential Games", *SIAM J. Control*, Vol. 18, no. 1, pp. 33-48. 19, no. , pp.

NEWTON-TYPE METHODS FOR STOCHASTIC GAMES

Boleslaw Tolwinski
Department of Mathematics, Colorado School of Mines
Golden, Colorado 80401

1. Introduction

Since the early books by Bellman (1957) and Howard (1960), it has been known that policy iteration is one of the most efficient procedures for solving Markov decision problems. In 1969 Pollatschek and Avi-Itzhak proposed to apply policy iteration to the solution of zero-sum stochastic games and showed that their approach was equivalent to solving the dynamic programming optimality equation by the classical Newton-Raphson technique. The Newton method representation of policy iteration was later explored in the context of Markov decision problems by Puterman and Brumelle (1978), and by Puterman and Shin (1978) who proposed a modification of the policy iteration algorithm equivalent to replacing the ordinary Newton iteration by the iteration of a quasi-Newton type.

The policy iteration algorithm as well as the method of Puterman and Shin are guaranteed to converge for any discounted Markov decision problem. Unfortunatly, the same is not true for stochastic games, as pointed out by Van Der Wal (1978). Motivated by this fact, Filar and Tolwinski (1987) have recently proposed a Modified Newton's Method (MNM, for short) which is guaranteed to converge for any discounted zero-sum stochastic game with finite state and action spaces. This paper presents the results of numerical experimentation with MNM and compares it with its quasi-Newton version (QNM) obtained by combining the scheme of Filar and Tolwinski (1987) with the ideas of Puterman and Shin (1978).

The plan of this paper is as follows. Section 2 includes the definition of a discounted stochastic game and the formulation of two problems associated with it, one belonging to the class of fixed-point problems and the other to the class of unconstrained optimization problems. In Section 3 the Modified Newton's Method and the Quasi-Newton Method are presented and the convergence of the latter is discussed. Finally, Section 4 is devoted to the discussion of the results of numerical experiments involving MNM and QNM.

2. Notation and Preliminaries

For our purposes, a finite state, finite action space stochastic game is defined by: a state space $S=\{1,2,\ldots,n\}$; for each $s \in S$, finite action spaces $A(s)=\{1,2,\ldots,m_s\}$ for player I and $B(s)=\{1,2,\ldots,n_s\}$ for player II; a reward law $r=\{r(s,i,j) \mid (s,i,j) \in S \times A(s) \times B(s)\}$; and a law of motion $p=\{p(t \mid s,i,j) \mid (s,i,j,t) \in S \times A(s) \times B(s) \times S\}$.

The game is played in stages $\tau=0,1,2,\ldots$. At some stage τ, the players find themselves in a state $s\varepsilon S$ and independently choose actions $i\varepsilon A(s)$, $j\varepsilon B(s)$. Player II pays player I the amount $r(s,i,j)$, and at stage $\tau+1$ the new state is $t\varepsilon S$ with probability $p(t|s,i,j)$, where $\Sigma_t p(t|s,i,j)=1$. Play continues at this new state.

In general, strategies can depend on complete histories of the game up until the current stage. We are concerned, however, with the simpler class of stationary strategies which depend only on the current state s. So, for player I, a stationary strategy x can be denoted by

$$x=(x(1),x(2),\ldots,x(n)),$$

where

$$x(s)=(x_1(s),x_2(s),\ldots,x_{m_s}(s))$$

and $x_i(s)$ denotes the probability that player I chooses action $i\varepsilon A(s)$ whenever the game is in state $s\varepsilon S$. Similarly, a stationary strategy y for player II can be defined as a set of n vectors y(s), each of which is a randomization on B(s). X and Y will denote the sets of all stationary strategies for players I and II respectively.

Once a pair of strategies (x,y) and an initial state $s\varepsilon S$ are fixed, the stream of expected payoffs to player I at stage τ, denoted by $V_\tau(s,x,y)$, $\tau=0,1,2,\ldots$, is well defined as is the resulting discounted game payoff

$$\Phi(s,x,y)=\sum_{\tau=0}^{\infty}\beta^\tau V_\tau(s,x,y),$$

where $\beta\varepsilon(0,1)$ is the fixed discount factor.

A pair of strategies (x^o,y^o) is optimal for players I and II if for all $s\varepsilon S$

$$\Phi(s,x,y^o)\leq\Phi(s,x^o,y^o)\leq\Phi(s,x^o,y),$$

for all strategies x and y of players I and II. In such a case, the number $v^*(s)=\Phi(s,x^o,y^o)$ is called the value of the game starting in state s, and $v^*=(v^*(1),\ldots,v^*(n))^T$ is called the value vector.

The well known theorem by Shapley (1953) asserts the existence of optimal stationary strategies for a class of stochastic games slightly more general than the discounted stochastic games defined above. The proof of that theorem is based on the following construction.

Let $v=(v(1),v(2),\ldots,v(n))^T$ be an arbitrary vector and for each $(s,i,j)\varepsilon S\times A(s)\times B(s)$ define

$$h(v,s,i,j)=r(s,i,j)+\beta \sum_{t=1}^{n} p(t|s,i,j)v(t).$$

We can now define a set of n matrix games, of dimensions $m_s \times n_s$, respectively

$$H(v,s)=(h(v,s,i,j)).$$

Let the value of each of these be denoted by

$$T(v,s):=val[H(v,s)] \qquad (2.1)$$

for each $s \in S$, and let

$$T(v)=(T(v,1),T(v,2),\dots,T(v,n))^T.$$

Now T represents a contraction mapping on R^n. Shapley proved that v^*, the value of the stochastic game, is the fixed point of T, or equivalently, that it is the unique root of the equation

$$\Psi(v):=T(v)-v=0 \qquad (2.2)$$

We shall now need additional notation: Let $x(v,s)$ and $y(v,s)$ be optimal strategies for players I and II, respectively, in the matrix games $H(v,s)$, for each $s \in S$. Note that these define the stationary strategy in the stochastic game,

$$x(v)=(x(v,1),\dots,x(v,n))$$

for player I, and a similarly defined stationary strategy $y(v)$ for player II. Furthermore, for any stationary strategy pair $(x,y) \in X \times Y$ we define the following quantities:

$$r(s,x,y) := \Sigma_i \Sigma_j \ r(s,i,j)x_i(s)y_j(s), \ s \in S;$$

$$r(x,y) := (r(1,x,y),\dots,r(n,x,y))^T;$$

$$p(t|s,x,y) := \Sigma_i \Sigma_j \ p(t|s,i,j)x_i(s)y_j(s), \ s,t \in S;$$

and the transition matrix of a Markov chain on the states of the game induced by (x,y), namely

$$P(x,y) := (p(t|s,x,y))_{s,t=1}^{n,n}$$

It is easy to verify that for each $s \in S$

$$\Phi(s,x,y)=[(I-\beta P(x,y))^{-1} r(x,y)]_s, \qquad (2.3)$$

where $[u]_s$ denotes the s-th entry of a vector u. Let $\Phi(x,y)$ denote the n-dimensional vector whose s-th entry is $\Phi(s,x,y)$. Now, with $(x(v),y(v)) \in X \times Y$ defined as above we can associate the Markov matrix

$$P(v) := P(x(v),y(v)).$$

Suppose that v^k is the current estimate of v^*, the value of the stochastic game. Then the policy iteration can be written as

$$v^{k+1}=\Phi(x(v^k),y(v^k)) \qquad (2.4)$$

To establish the equivalence of (2.4) with Newton's method one needs the following theorem due to Pollatschek and Avi-Itzhak (1969).

Theorem 2.1

(i) The partial derivatives of

$$\Psi(v,s) := T(v,s)-v(s)$$

are unique and continuous in v almost everywhere, for each $s \in S$.

(ii) If for each $s \in S$, $x(v,s)$ and $y(v,s)$ are basic optimal strategies in the matrix game $H(v,s)$, in the sense of Shapley and Snow (1950), then whenever it is unique

$$\partial\Psi(v,s)/\partial v(t)=\beta p(t|s,x(v),y(v))-\delta(s,t),$$

where $\delta(s,t)$ is the Kronecker delta.

(iii) In view of (ii), the Jacobian matrix of $\Psi(v)$ satisfies:

$$\Psi'(v)=\beta P(v)-I$$

almost everywhere.

Note that since $P(v)$ is a Markov matrix and $\beta \in (0,1)$ it follows that the inverse of the Jacobian

$$[\Psi'(v)]^{-1}=-[I-\beta P(v)]^{-1} \qquad (2.5)$$

is well defined almost everywhere. Consequently, the classical Newton iteration for the solution of (2.2) can be written, with the help of (2.1) and (2.3) as

$$\begin{aligned} v^{k+1} &= v^k-[\Psi'(v^k)]^{-1}\Psi(v^k) \\ &= v^k+[I-\beta P(v^k)]^{-1}[T(v^k)-v^k] \\ &= v^k+[I-\beta P(v^k)]^{-1}[r(x(v^k),y(v^k)) \\ &\qquad +\beta P(v^k)v^k-v^k] \\ &= [I-\beta P(v^k)]^{-1}r(x(v^k),y(v^k)) \\ &= \Phi(x(v^k),y(v^k)) \end{aligned} \qquad (2.6)$$

where v^k is the current estimate of the solution of (2.2). It is thus clear that policy iteration is equivalent to Newton's method for the discounted zero-sum stochastic games defined earlier in this section.

Since the example of Van Der Wal shows that policy iteration as defined by (2.4) need not converge, the following modification of Newton's method has been proposed in (Filar and Tolwinski, 1987). Define

$$J(v) := (1/2)[\Psi(v)]^T\Psi(v) \qquad (2.7)$$

for all v in R^n, and consider the unconstrained minimization problem:

$$\min[J(v)]. \qquad (2.8)$$

Note that that the gradient (row) vector of J(v) is given by

$$J'(v)=[\Psi(v)]^T[\Psi'(v)] \qquad (2.9)$$

and that $J'(v)=0$ if and only if $\Psi(v)=0$ (since the matrix $\Psi'(v)$ is nonsingular by Theorem 2.1 (iii)). The new iterative step is now defined by the formula

$$v^{k+1}=v^k-w^k[\Psi'(v^k)]^{-1}\Psi(v^k), \qquad (2.10)$$

where the "step-size" w^k is selected so as to ensure that

$$J(v^{k+1})<J(v^k). \qquad (2.11)$$

Note that the search direction in (2.10), i.e.,

$$s^k=-[\Psi'(v^k)]^{-1}\Psi(v^k) \qquad (2.12)$$

is in fact a direction of descent for J(v) because one has

$$J'(v^k)s^k=-\|\Psi(v^k)\|_2^2<0 \qquad (2.13)$$

with the strict inequality holding for any v^k which is not a solution to the equation (2.1).

3. Two Newton-Type Algorithms for Stochastic Games

The discussion at the end of the previous section leads to the following algorithm:

Algorithm I (Modified Newton's Method, MNM)

Step 0. Set k:=0. Select two parameter values: $\alpha\in(0,1)$ ("small" value of α is generally recommended), and $\mu\in[0.5,0.8]$. Also select v^o, the initial estimate of the value vector.

Step 1. Calculate for each $s\in S$, the matrix game $H(v^k,s)$, a pair of optimal basic strategies $(x(v^k,s),y(v^k,s))$ for players

I and II (see Goldman and Tucker, 1956), and its value $T(v^k,s)$. Calculate $T(v^k)$, $\Psi(v^k)$ and $J(v^k)$ (see (2.1), (2.2) and (2.7)).

Step 2. If $J(v^k)=0$, stop; v^k is the value vector of the stochastic game.

Step 3. Calculate the transition matrix $P(v^k)$ and determine s^k by solving the system of linear equations

$$(I-\beta P(v^k))s^k=\Psi(v^k) \tag{3.1}$$

(recall that $(I-\beta P(v^k))=-\Psi'(v^k)$)

Step 4. Set $w:=1$.

Step 5. Test the inequality

$$J(v^k+ws^k)-J(v^k)\leq\alpha wJ'(v^k)s^k. \tag{3.2}$$

If the above inequality is satisfied set $v^{k+1}:=v^k+ws^k$, $k:=k+1$ and return to Step 2. Otherwise go to Step 6.

Step 6. Set $w:=\mu w$ and return to Step 5.

Comment: The objective of Steps 5 and 6 is to select the step-size w according to the well known Armijo's rule e.g., see (McCormick, 1983)).

Theorem 3.1

For any choice of the initial estimate v^o, the sequence $\{v^k\}$ generated by MNM converges to the limit v^* which is the value of the stochastic game.

Proof. (see (Filar and Tolwinski, 1987)).

One of the costliest operations in Algorithm I is the computation of s^k by the solution of the system (3.1). The way to avoid solving (3.1) at each iteration is to consider some approximation of the inverse of the matrix $\Psi'(v^k)$. The approach which turns out to be quite successful is to use the Neumann series representation of $(I-\beta P)^{-1}$, i.e., the expression

$$-[\Psi'(v)]^{-1}=[I-\beta P(v)]^{-1}=\sum_{i=0}^{\infty}(\beta P(v))^i \tag{3.3}$$

Obviously, the matrix $-[\Psi'(v)]^{-1}$ can be approximated with arbitrary precision by matrices $Q_m(v)$, where

$$Q_m(v)=\sum_{i=0}^{m}(\beta P(v))^i \tag{3.4}$$

and consequently, the iteration (2.10) can be replaced by

$$v^{k+1}=v^k+w^kQ_m(v^k)\Psi(v^k), \tag{3.5}$$

where $s^k=Q_m(v^k)\Psi(v^k)$ defines the new search direction.

The expression (3.4) would be very inefficient as the way to compute Q_m and s^k. However, following the reasonning similar to that given in (Puterman and Shin, 1978), one can establish the following relations:

$$u:=v^k+Q_m(v^k)\Psi(v^k)$$

$$=v^k+\sum_{i=0}^{m}(\beta P(v^k))^i\Psi(v^k)$$

$$=v^k+\sum_i(\beta P(v^k))^i[r(x(v^k),y(v^k))+(\beta P(v^k)-I)v^k]$$

$$=v^k+\sum_i(\beta P(v^k))^i r(x(v^k),y(v^k))+\sum_i(\beta P(v^k))^{i+1}v^k-\sum_i(\beta P(v^k))^i v^k$$

$$=\sum_{i=0}^{m}(\beta P(v^k))^i r(x(v^k),y(v^k)) + (\beta P(v^k))^{m+1}v^k$$

$$=T_k^{m+1}(v^k), \tag{3.6}$$

where the superscript indicates operator composition and T_k is defined by

$$T_k(v)=r(x(v^k),y(v^k))+\beta P(v^k)v \tag{3.7}$$

Thus, the new expression for the search direction is

$$s^k=u-v^k=T_k^{m+1}(v^k)-v^k, \tag{3.8}$$

and this expression provides a relatively efficient way for the computation of s^k.

On the basis of the above discussion one can define the following algorithm:

<u>Algorithm II</u> (Quasi-Newton Method, QNM)

<u>Step 0.</u> Select parameters $\alpha\epsilon(0,1)$, $\mu\epsilon[0.5,.8]$, $\delta\epsilon(0,\delta^*]$, $m_0\geq 1$. Also select v^0, the initial estimate of the value vector. Set $k:=0$ and $m:=m_0$.

<u>Comment</u>: δ^* may be problem dependent. See the discussion below.

Step 1. Calculate for each $s \in S$, the matrix game $H(v^k,s)$, a pair of optimal basic strategies $(x(v^k,s),y(v^k,s))$ for players I and II, and the value $T(v^k,s)$. Calculate $T(v^k)$, $\Psi(v^k)$, and $J(v^k)$.

Step 2. If $J(v^k)=0$, stop; v^k is the value vector of the stochastic game.

Step 3. Calculate the search direction $s^k = T_k{}^m(v^k)-v^k$.

Step 4. Test the inequality

$$J'(v^k)s^k/(\|J'(v^k)\|\ \|s^k\|) \leq -\delta \qquad (3.9)$$

If the above inequality is satisfied, go to Step 5. Otherwise, set $m:=m+1$ and return to step 3.

Comment: Step 4 ensures that s^k is a descent direction for $J(v)$.

Step 5. Set $w:=0$.

Step 6. Test the inequality (3.2). If it is satisfied set $v^{k+1}:=v^k+ws^k$, $k:=k+1$ and return to Step 2. Otherwise go to Step 7.

Step 7. Set $w:=\mu w$ and return to Step 6.

Theorem 3.2

For a given stochastic game there exists a positive number δ such that the sequence $\{v^k\}$ generated by QNM converges to v^*, the value of the stochastic game, for any initial estimate v^o. In addition, given a sufficiently small δ, the value of m remains bounded by a number depending only on the discount parameter β and on n, the cardinality of the state space.

To prove Theorem 3.2 we shall need the following lemmas:

Lemma 3.3

For the minimization problem (2.8), let s^k be a search direction, let w^k be a step-size chosen according to Armijo's rule, and assume that the following hold:

(a) There exist scalars a, M >0 such that for all k=0,1,2,...,

$$a\|J'(v^k)\|_2 \leq \|s^k\|_2 \leq M.$$

(b) There exists $\delta>0$ such that for all k for which $J'(v^k)\neq 0$

$$J'(v^k)s^k/(\|J'(v^k)\|_2\|s^k\|_2)\leq-\delta<0$$

Define the sequence $\{v^k\}$ by the formula $v^{k+1}=v^k+w^k s^k$ and suppose that u is an accumulation point of that sequence. Then

$$J'(u)=0.$$

Proof. See Theorem 12 on page 137 of McCormick (1983).

<u>Lemma 3.4</u>

Consider the minimization problem (2.8) induced by the discounted stochastic game, and let s^k be given by (3.8). Then, condition (a) of Lemma 3.3 is always satisfied and condition (b) holds with

$$\delta=(1-\beta^{m+1}c_1)(1-\beta)/[c_1c_2(1+\beta)] \tag{3.10}$$

where $c_1=n^{1/2}$, $c_2=n^{3/2}$, and $\delta>0$ for sufficiently large m.

Proof.

Note first that

$$\|\Psi(v^k)\|_2=\|[I-\beta P(v^k)]\|_2\leq c_1\|[I-\beta P(v^k)]\|_\infty \tag{3.11}$$

where $c_1=n^{1/2}$. On the other hand,

$$\|[I-\beta P(v^k)]\|_\infty=\max_s \Sigma_t|\beta p(t|s,x(v^k),y(v^k))-\delta(s,t)| \tag{3.12}$$

and since for every $s\in S$

$$\Sigma_t|\beta p(t|s,\cdot,\cdot)-\delta(s,t)|=1+\beta[1-2p(s|s,\cdot,\cdot)] \tag{3.13}$$

one obtains that for every k=0,1,2,...

$$(1-\beta)\leq\|\Psi'(v^k)\|_\infty\leq(1+\beta) \tag{3.14}$$

and

$$\|\Psi'(v^k)\|_2\leq c_1(1+\beta) \tag{3.15}$$

Also, since $P(v^k)$ is a Markov matrix

$$\|(\beta P(v^k))^i\|_\infty=\beta^i \tag{3.16}$$

which implies that

$$\|Q_m(v^k)\|_2\leq c_1\|Q_m(v^k)\|_\infty\leq\sum_{i=0}^{m}\beta^i=(1-\beta^{m+1})/(1-\beta)\leq 1/(1-\beta) \tag{3.17}$$

$$\|s^k\|_2\leq\|Q_m(v^k)\|_2\|\Psi(v^k)\|_2\leq c_1\|\Psi(v^k)\|_2/(1-\beta) \tag{3.18}$$

and

$$\|[I-(\beta P(v^k))^{m+1}]^{-1}\|_2 \le c_1\|[I-(\beta P(v^k))^{m+1}]^{-1}\|_\infty$$

$$=c_1\|\sum_{i=0}^{\infty}(\beta P(v^k))^{(m+1)i}\|_\infty \le c_1/(1-\beta^{m+1}) \le c_1/(1-\beta) \qquad (3.19)$$

Finally, in a manner simalar to that used to obtain (3.14), one can verify that

$$\|[\Psi'(v^k)]^T\|_2 \le c_2(1+\beta), \qquad (3.20)$$

where $c_2=n^{3/2}$.

In view of the above inequalities the proof of the lemma becomes straightforward. Since the matrix $Q_m(v^k)$ can also be represented in the form

$$Q_m(v^k)=[I-\beta P(v^k)]^{-1}[I-(\beta P(v^k))^{m+1}], \qquad (3.21)$$

one can write

$$J'(v^k)^T=\Psi'(v^k)^T\Psi(v^k)=\Psi'(v^k)^TQ_m(v^k)^{-1}Q_m(v^k)\Psi(v^k)$$
$$=\Psi'(v^k)^TQ_m(v^k)^{-1}s^k \qquad (3.22)$$

and in view of (3.14), (3.19) and (3.20) conclude that

$$\|J'(v^k)\|_2 \le \|\Psi'(v^k)^T\|_2\|[I-(\beta P(v^k))^{m+1}]^{-1}\|_2\|I-\beta P(v^k)\|_2\|s^k\|_2$$
$$\le c_2(1+\beta)c_1(1+\beta)(c_1/(1-\beta))\|s^k\|_2 \qquad (3.23)$$

which is equivalent to the first inequality in condition (a) of Lemma 3.3. The second inequality follows from (3.18) and the fact that

$$\|\Psi(v^k)\|_2=(2J(v^k))^{1/2} \le (2J(v^o))^{1/2}, \text{ for all } k. \qquad (3.24)$$

To verify condition (b) note that

$$J'(v^k)s^k=\Psi(v^k)^T[\beta P(v^k)-I][I-\beta P(v^k)]^{-1}[I-(\beta P(v^k))^{m+1}]\Psi(v^k)$$
$$=-\Psi(v^k)^T[I-(\beta P(v^k))^{m+1}]\Psi(v^k)$$
$$=-\|\Psi(v^k)\|_2^2+\beta^{m+1}\Psi(v^k)^TP(v^k)^{m+1}\Psi(v^k)$$
$$\le\|\Psi(v^k)\|_2^2(-1+\beta^{m+1}\|P(v^k)^{m+1}\|_2)$$
$$\le\|\Psi(v^k)\|_2^2(-1+\beta^{m+1}c_1) \qquad (3.25)$$

therefore, for m such that $\beta^{m+1}c_1-1<0$

$$J'(v^k)s^k/(\|J'(v^k)\|_2\|s^k\|)$$
$$\le(\beta^{m+1}c_1-1)\|\Psi(v^k)\|_2^2/(\|\Psi'(v^k)^T\Psi(v^k)\|_2\|s^k\|)$$

$$\leq (\beta^{m+1}c_1-1)\|\Psi(v^k)\|_2^2/[\|\Psi'(v^k)^T\|_2\|\Psi(v^k)\|_2^2(c_1/(1-\beta))]$$

$$\leq (\beta^{m+1}c_1-1)(1-\beta)/[(1+\beta)c_1c_2]=-\delta \qquad (3.26)$$

This completes the proof of the lemma.

<u>Lemma 5.5</u>

The sequence $\{v^k\}$ generated by Algorithm II is bounded for all $k=0,1,2,\ldots.$

Proof.

Let $M=\max|r(s,i,j)|$ and consider the following relations:

$$v^{k+1} = v^k+w^ks^k = v^k+w^kQ_m(v^k)\Psi(v^k)$$

$$=v^k+w^k[I-(\beta P(v^k))^{m+1}][I-\beta P(v^k)]^{-1}\Psi(v^k)$$

$$=v^k+w^k\{[I-\beta P(v^k)]^{-1}\Psi(v^k)-(\beta P(v^k))^{m+1}[I-\beta P(v^k)]^{-1}\Psi(v^k)\} \qquad (3.27)$$

and

$$[I-\beta P(v^k)]^{-1}\Psi(v^k)=[I-\beta P(v^k)]^{-1}\{r[x(v^k),y(v^k)]-[I-\beta P(v^k)]v^k\}$$

$$=\Phi[x(v^k),y(v^k)]-v^k \qquad (3.28)$$

Combining (3.27) and (3.28) one obtains

$$v^{k+1}=(1-w^k)v^k+w^k\{\Phi[x(v^k),y(v^k)]-(\beta P(v^k))^{m+1}[I-\beta P(v^k)]^{-1}\Psi(v^k)\} \qquad (3.29)$$

and

$$\|v^{k+1}\|_2\leq (1-w^k)\|v^k\|_2+w^k(\|\Phi[x(v^k),y(v^k)]\|_2+\|(\beta P(v^k))^{m+1}\|_2\|\Psi(v^k)\|_2)$$

$$\leq (1-w^k)\|v^k\|+w^k[M/(1-\beta)+c_1\beta^{m+1}(2J(v^o))^{1/2}] \qquad (3.30)$$

Now, let $N:=M/(1-\beta)+c_1\beta(2J(v^o))^{1/2}$ and assume (without loss generality) that $\|v^o\|_2\leq N$. Since w^k is a number from the interval $(0,1]$, (3.30) implies that

$$\|v^k\|_2\leq N \text{ for all } k=0,1,2,\ldots \qquad (3.30)$$

which completes the proof.

<u>Proof of Theorem 3.2</u>

By Lemma 3.5 the sequence $\{v^k\}$ has at least one accumulation point, say u. By lemmas 3.3 and 3.4 we must have $J'(u)=0$ and therefore also $\Psi(u)=0$. Consequently, u is the value vector of the stochastic game. By the uniqueness of the letter, $\{v^k\}$ can have only one accumulation point. Thus, $\{v^k\}$ has exactly one accumulation point which is the value of the game. This completes the proof of the theorem.

4. <u>Results of numerical experiments</u>

In order to test and compare the two algorithms defined in the previous section a number of numerical experiments have been conducted on the Colorado School of Mines VAX 8600 computer. About thirty randomly generated test problems were considered, each for at least three values of the discount parameter β. The initial value for each example was $v^0=0$, and the stopping criterion was $J(v^k)<.0000005$. In the tables given below each test problem is characterized by the following data: l - the number of actions (assumed identical in each state and for each player); n - the number of states; and k which is such that for every fixed triple (s,i,j), there are at most k strictly positive transition probabilities $p(t|s,i,j)$ for $t \in S$.

In Table 1 the number of iterations necessary to achieve the solution (with the given precision) with help of Algorithm I (MNM) is given for three values of β, namely $\beta=0.8$, 0.9, and 0.95. For example, the entry (3,4,3) in row (200,20) and column 4 means that for the randomly generated stochastic game with 4 actions, 200 states, and at most 20 strictly positive transition probabilities in each state, Algorithm I required 3, 4, and 3 iterations to achieve the solution for $\beta=0.8$, 0.9, and 0.95 respectively. The results presented in Table 1 seem to indicate that, at least in the considered range of parameters l and n, the number of iterations increases very slowly with n and actually decreases when the number of actions grows.

(n,k) \ l	2	4	5	10	20
(100,2)	(4,5,5)	(3,4,4)	(3,3,3)	(2,2,2)	(2,2,2)
(100,4)	(4,5,4)	(3,3,3)	(3,3,3)	(2,2,2)	(2,2,2)
(100,10)	(4,4,5)	(3,3,3)	(3,3,3)	(2,2,2)	
(200,2)	(5,6,5)	(4,5,4)		(2,2,2)	(2,2,2)
(200,4)	(4,4,4)	(3,4,4)	(3,3,3)		
(200,20)	(4,5,5)	(3,4,3)	(3,3,3)		
(200,100)	(3,5,5)				
(400,2)	(5,5,5)	(4,4,5)			
(600,2)	(5,5,6)				

Table 1. Number of iterations for MNM.

Table 2 gives the execution times per iteration, expressed in seconds of CPU-time, for Algorithm I. Since these times turned out to be almost identical for all values of β (the differences were well within the margin of error of the CPU-timer employed), only execution times for β=0.8 have been given.

l / (n,k)	2	4	5	10	20
(100,2)	(0.85)	(2.15)	(3.10)	(10.50)	(54.00)
(100,4)	(1.10)	(3.25)	(4.80)	(16.80)	(79.00)
(100,10)	(2.10)	(6.30)	(9.30)	(35.50)	
(200,2)	(3.60)	(9.50)		(42.00)	(186.00)
(200,4)	(6.00)	(14.60)	(20.00)		
(200,20)	(16.20)	(46.20)	(68.50)		
(200,100)	(56.25)				
(400,2)	(20.00)	(52.00)			
(600,2)	(56.00)				

Table 2. CPU-time per iteration in seconds for MNM.

Table 3 concerns Algorithm II and gives the optimal values of the parameter m for the three values of β, e.g., the entry (20,45,60) in row (100,4) and column 10 means that Algorithm II computed the solutions of the corresponding stochastic games with the discount parameter equal to 0.8, 0.9, and 0.95 in the shortest time when the parameter m_o in Step 0 of the algorithm was assigned values 20, 45, and 60 respectively (afterwards, m stayed equal to m_o until the computations where completed). As expected, m turned out to strongly depend on the discount parameter β, but also apeared to increase with the number of actions and seemed to be rather insensitive to the number of states. In virtually all cases, the given value of m turned out to be optimal when Algorithm II was able to find the solution of the game in exactly the same number of iterations as Algorithm I. The parameter m of a smaller magnitude which led to a larger number of iterations usually resulted in longer execution times.

(n,k) \ 1	2	4	5	10	20
(100,2)	(10,20,35)	(10,20,40)	(15,30,55)	(20,40,80)	(20,40,80)
(100,4)	(15,20,45)	(15,30,55)	(20,35,55)	(20,45,60)	(20,40,80)
(100,10)	(10,25,35)	(15,30,55)	(15,30,60)	(20,40,80)	
(200,2)	(10,15,35)	(10,20,45)		(20,50,100)	(20,55,110)
(200,4)	(10,25,45)	(15,25,45)	(15,30,60)		
(200,20)	(10,20,35)	(15,25,50)	(15,30,60)		
(200,100)	(20,20,45)				
(400,2)	(10,20,40)	(15,30,35)			
(600,2)	10,20,40)				

Table 3. Optimal values of m for QNM.

Table 4 gives the execution times per iteration for Algorithm II. Only the results for $\beta=0.8$ and 0.9 are presented, since it turned out that for $\beta=0.95$ Algorithm I rather consistently outperformed Algorithm II. Unlike the performance of MNM, the performance of QNM strongly depends on the value of the discount parameter and deteriorates when β approaches one.

(n,k) \ 1	2	4	5	10	20
(100,2)	(0.6,0.8)	(1.8,1.9)	(2.6,2.9)	(10.0,10.7)	(53.9,54.2)
(100,4)	(0.8,1.0)	(2.7,3.0)	(4.0,4.5)	(15.9,16.1)	(78.0,79.0)
(100,10)	(1.5,1.8)	(5.3,5.8)	(8.2,8.6)	(33.5,33.7)	
(200,2)	(2.7,3.2)	(6.8,7.7)		(37.9,40.6)	(182.,185.)
(200,4)	(3.7,5.1)	(10.9,11.9)	(16.0,17.7)		
(200,20)	(10.4,11.3)	(38.4,39.4)	(60.7,61.8)		
(200,100)	(38.4,38.5)				
(400,2)	(13.1,20.0)				
(600,2)	(44.8,66.3)				

Table 4. CPU-time per iteration for QNM.

Finally, Table 5 compares the performance of the two algorithms. It shows that the improvement in the performance of Algorithm II as compared with Algorithm I depends not only on β but also on the number of actions. It can be easily explained by the fact that for games with larger action spaces most of the time is spent on solving matrix games $H(v^k,s)$ and improvements in the efficiency of the computation of s^k become less relevant to the overall execution time.

(n,k) \ l	2	4	5	10	20
(100,2)	(26,7)	(15,13)	(15,6)	(4,-)	(-,-)
(100,4)	(28,9)	(17,9)	(16,6)	(7,4)	(-,-)
(100,10)	(30,16)	(17,8)	(12,9)	(4,-)	
(200,2)	(24,11)	(28,20)		(10,3)	(2,-)
(200,4)	(25,19)	(22,14)	(22,14)		
(200,20)	(37,30)	(17,14)	(11,10)		
(200,100)	(32,32)				
(400,2)	(36,-)	(39,21)			
(600,2)	(20,-)				

Table 5. Savings in execution time when using QNM as compared with MNM, expressed in terms of $[(t_N-t_Q)/t_N]\times 100\%$, where t_N and t_Q represent the execution times of MNM and QNM respectively. Symbol "-" corresponds to the case when the performence of QNM was inferior to the performance of MNM.

The empirical evidence presented recently by Breton (1987) and Breton et al. (1986) suggested that policy iteration was the fastest method for solving stochastic games in cases when it converged. The Modified Newton Method is guaranteed to converge for any stochastic game and any initial estimate v^0, while in most cases the computations under MNM are identical to those under policy iteration implying, therefore, that MNM will usually have the same rate of convergence as policy iteration has. The Quasi-Newton Method has the same convergence properties as MNM but is up to 40 percent faster than the latter for problems where the number of states is large relative to number of actions and the discount parameter does not exceed 0.9. Additional numerical experimentation should provide more information about optimal values of the parameter m as a function of the discount parameter and the cardinality of the state space versus cardinalities of the action spaces, thus making QNM a practical tool for the solution of discounted stochastic games.

5. Conclusions and Open Problems

The two algorithms discussed in this paper permit one to solve zero-sum discounted stochastic games of the size which might be large enough to make stochastic game models relevant to practical applications. The approach which led to the development of MNM can probably be extended to zero-sum stochastic games under the limit of averages criterion. Another challenging direction of research is the development of efficient algorithms for nonzero-sum stochastic games in both the discounted and limiting average cases. Empirical evidence indicates that policy iteration quite often fails to converge when applied to nonzero-sum games. Indeed, in that context policy iteration is no longer equivalent to Newton's method, and Newton's method itself is not directly applicable because the derivative of the appropriate value operator cannot be explicitly calculated even if it exists. The results of some preliminary numerical experimentation indicate that policy iteration modified along the lines of Algorithm I is more likely to converge to a solution of a nonzero-sum game than policy iteration in its original form. Nevertheless, the convergence is by no means guaranteed and the issue is far from being closed.

References

Bellman, R.E. (1957), Dynamic Programming, Princeton University Press, Princeton N.J.

Breton, M. (1987), Equilibre Pour Des Jeux Sequentiel, PhD thesis, University of Montreal.

Breton, M., J.A. Filar, A. Haurie, and T.A. Schultz (1986) "On the computation of equilibria in discounted stochastic games", in Dynamic Games and Applications in Economics, T. Basar (ED.), Springer Verlag. Lecture Notes in Economics and Mathematical Systems.

Filar, J.A., and B. Tolwinski (1987), "On the algorithm of Pollatschek and Avi-Itzhak", (submitted for publication).

Goldman, A.J., and A.W. Tucker (1956), "Theory of linear programming", in Linear Inequalities and Related Systems, H.W. Kuhn and A.W. Tucker (Eds.), Princeton University Press, Annals of Mathematics Studies #38.

Howard, R.A. (1960), Dynamic Programming and Markov Processes, The MIT Press, Cambridge, MA.

McCormick, G.P. (1983), Nonlinear Programming, Wiley, New York.

Pollatschek, M. and B. Avi-Itzhak (1969), "Algorithms for stochastic games with geometric interpretation", Management Science, 15, 399-415.

Puterman, M.L. and S.L. Brumelle (1978), "The analytic theory of policy iteration", in Dynamic Programming and Its Applications, M.L. Puterman (Ed.), Academic Press, New York.

Puterman, M.L. and M.C. Shin (1978), "Modified policy iteration algorithms for discounted Markov decision problems", Management Science, 24, 1127-1137.

Shapley, L.S. (1953), "Stochastic games", Proceedings of the National Academy of Science USA, 39, 1095-1100.

Shapley, L.S. and R.N. Snow (1950), "Basic solutions of descrete games", in Contributions to the Theory of Games, H.W. Kuhn and A.W. Tucker (Eds.), Princeton University Press, Annals of Mathematics Studies #24.

Van Der Wal, J. (1978), "Discounted Markov games: generalized policy iteration method", J. Optimiz. Th. Appls., 25, 125-138.

NASH-COURNOT EQUILIBRIA IN THE EUROPEAN GAS MARKET: A CASE WHERE OPEN-LOOP AND FEEDBACK SOLUTIONS COINCIDE

Sjur Didrik Flåm
Institute of Economics, University of Bergen, Norway

Georges Zaccour
GERAD, Ecole des H.E.C., Montreal, Canada and
INSEAD, Fontainebleau, France

Abstract

We consider an oligopolistic industry extracting a non-renewable resource sold in a competitive market. We show, first, that if all players, but one, have infinite private reserves and production capacities (or alternatively the expansion of the production capacities is exogeneous) then Nash open-loop and feedback equilibria coincide. Second, we show that these conditions correspond to a fairly good approximation of the European gas market and hence the practical importance of our result.

1. Introduction

The purpose of this note is two-fold. First, we show that, under some conditions, Nash-Cournot equilibria in open-loop and feedback strategies coincide. Second, the practical importance of our result is illustrated in the case of the European gas market.

When one assumes that all players (firms) make their decisions simultaneously without any cooperation, then the natural equilibrium concept in static oligopoly is that of Nash-Cournot. This equilibrium corresponds to a state of the game where no player finds it advantageous to change its decision unilaterally, given the choices of other players.

In a dynamic context, there exists several ways for defining Nash equilibrium depending on the type of information that players consider when making their decisions. The information structure in open-loop corresponds to the case where they utilize only <u>time</u> as information on which they base their choices

(1) The second author has been supported by CRSNG - Canada

whereas the feedback information structure corresponds to the case where players utilize both time and state variables as information. It is well known (see Newberry (1984)) that the open-loop equilibrium is not perfect a la Selten (1975)) and thus may be dynamically inconsistent; i.e., if one or several players do not play according to their equilibrium solution for some time, and if one recomputes, at a subsequent period τ the equilibrium strategies for all players, this equilibrium may not correspond to the one initially computed for the periods $T-\tau$, where T is the terminal date of the game. To avoid this possible dynamic inconsistency one has to state explicitly that the players are committed to follow their original plans. This can be done by interpreting the equilibrium trajectories as supply contracts that players are obliged to fulfil (Newberry (1984), Haurie et al (1987) b)). The Nash equilibrium in feedback strategies is by construction subgame perfect and dynamically consistent. However, it is usually very hard to compute. This explains why the open-loop solution is adopted in most real applications (see for instance Salant (1982) and Haurie et al (1987)b). For a detailed discussion of both equilibrium concepts in the context of non-renewable resources, the interested reader could refer to Reinganum and Stockey (1985), Newberry (1984) and Salant (1982) and to references therein.

In a stochastic and dynamic context, Haurie et al (1987a) suggested the concept of S-adapted open-loop equilibrium where the players base their decisions on time and the state of the world. This equilibrium lies "half-way" between the open-loop and the feedback concepts. It is not subgame perfect, however it inherits the computational simplicity of the open-loop equilibrium.

It appears from what has been said above that when one considers a real application of dynamic game theory, one faces the dilemma between choosing the credible hard to compute feedback solution and the non-self supporting but tractable open-loop equilibrium. Even if one states that open-loop trajectories are irrevocable supply contracts it is not clear how empirically one can be sure that they are going to be effectively fulfilled. When saying that we have in mind the European gas market where traditionally suppliers and buyers sign long-term contracts (and they are still insisting on doing so) which are renegotiated frequently. Actually, this market, to be described later on, has recently gone through a mania of contractual rearrangements and which is now institutionalizing price renegotiations (Haurie et al (1987b)). To avoid this dilemma authors have been particularly interested in open-loop equilibria that happen to be subgame perfect. In particular, Eswaran and Lewis (1985) showed that if the reserves are private, then Nash-Cournot equilibria in open-loop and feedback strategies coincide in the following cases:

(1) isoelastic demand and zero extraction cost

(2) linear demand, quadratic extraction cost and symmetric conditions.

Furthermore, Eswaran and Lewis noticed that the difference between the two solutions is very small in other situations. In all cases analyzed by these authors, the state equation is of the following type:

$$R_{i,t+1} = R_{i,t} - q_{i,t}$$

where $R_{i,t}$ and $q_{i,t}$ denote respectively reserves and production of player i at period t.

In this paper, we are interested in the particular case where only one player is constrained by such a state equation (it may be a relation involving reserves depletion or expansion of the production capacity or both), all other players own infinite reserves and production capacities (alternatively the expansion of production capacity is exogeneous). The assumptions made on demand and cost functions are merely those needed to prove existence of at least one dynamic Nash-Cournot equilibrium and consequently are more general than those adopted by Eswaran and Lewis. However we get the same result as these authors; i.e. under conditions stated above Nash-Cournot equilibria in open-loop and feedback strategies coincide. This result which is close to trivial from a theoretical point of view has however, practical importance since there exists markets, in particular the European gas one, which will serve as an illustration later on, where these conditions are fulfilled.

The remainder of the paper is organized as follows: in section 2, we present the problem of an oligopolistic industry extracting a non-renewable resource sold in a competitive market and prove our main result. In section 3, we give a brief description of the European gas market and show, only by referring to hypothesis accepted in the literature but arranged somewhat differently, that our conditions correspond to a fairly good approximation of the European gas market. Finally, section 4 is devoted to a very brief presentation of what has been published recently concerning the computation of Nash-Cournot equilibria that could eventually be applied in order to get numerical results.

2. A non-cooperative dynamic game

We consider in this section an oligopolistic industry extracting an homogeneous non-renewable resource and facing a competitive demand.

Let I be the set of producers. Each element i I is described by the following data:

(1) $R_{i,t}$, the available reserves at period $t = o,\ldots,T$, $R_{i,o}$ given.

(2) $K_{i,t}$, the installed production capacity at t, $K_{i,o}$ given.

(3) $q_{i,t}$, the production at period t.

(4) $C_i(q_{i,t})$: R+ -----> R+, the production cost function.

(5) $\Gamma_i(I_{i,t})$: R+ -----> R+, the investment cost function, where I_i denotes the physical capacity invested at period t.

(6) β_i, the discount factor.

The consumers are described by an inverse demand law $P_t(Q_t)$ where $Q_t = \sum_{i \in I} q_{i,t}$ and $P_t(\cdot)$ denotes the price consumers are willing to pay at period t if the total quantity Q_t is available on the market.

If one assumes that each producer maximizes the sum of its discounted profits of the T-period game, then he faces the following optimization problem:

$$\text{Max } \Pi_i = \sum_{t=o}^{T} \beta_i^t \{ q_{i,t} P_t(Q_t) - C_i (q_{i,t}) - \Gamma_i (I_{i,t}) \}$$

Subject to:

(1) Reserves depletion

$$R_{i,t+1} = R_{i,t} - q_{i,t} \qquad t = o, \ldots, T-1$$

(2) Expansion of production capacity

$$K_{i,t+\tau} = (1 - \delta_i) K_{i,t} + I_{i,\tau} \qquad t = o, \ldots, T-\tau$$

(3) Production capacity

$$q_{i,t} \leq K_{i,t} \qquad t = o, \ldots, T$$

(4) Non-negativity of control parameters and reserves

$$q_{i,t}, I_{i,t}, R_{i,t} \geq o \qquad t = o, \ldots, T$$

where δ denotes the physical depreciation rate of production capacity and τ the investment delay.

Let $A_{i,t}(R_{i,t}, K_{i,t})$ be the set of admissible actions of player i at period t. Denote by $y_{i,t}$ the pair of decision variables of player i at period t, e.g. $y_{i,t} = (q_{i,t}, I_{i,t})$, and let $y_i = (q_{i,t}, I_{i,t})_{t=o,\dots,T}$, $y = (y_1, \dots, y_I)$ and by $y_t = (y_{1,t}, \dots, y_{I,t})$.

Definition: $y^* = (y_1^*, \dots, y_I^*)$ is a dynamic Nash equilibrium in open-loop if $Y_i^*, \forall i \in I$, is an optimal solution of the following program:

$$\text{Max } \Pi_i = \sum_{t=o}^{T} \beta_i^t \{q_{i,t} \; P_t (q_{i,t} + Q_{i,t}^*) - C_i(q_{i,t}) - \Gamma_i(I_{i,t}) \}$$

subject to $y_{i,t} \in A_{i,t}(R_{i,t}, K_{i,t}), \forall t$

and where $Q_{i,t}^* = \sum_{\substack{j \in I \\ j \neq i}} q_{j,t}^*$.

Furthermore, if for any period $\tau \geq o$, $y_\tau^*, \dots, y_T^*$ is an equilibrium for the subgame starting at that period, with y_t^* ($\tau \leq t \leq T$) depending on the current state, then this trajectory is called a Nash equilibrium in feedback strategies.

Proposition: There exists at least one dynamic Nash equilibrium under the following assumptions:

(H1) $P_t(Q_t)$ and $Q_t P_t(Q_t)$ are strictly decreasing and continuously differentiable.

(H2) $C_i(.)$ and $\Gamma_i(.)$ are convex, increasing and continously differentiable.

Proof: Under (H1), Murphy et al (1982) showed that the revenue of each player $q_{i,t} P_t(.)$ is a strictly concave function with respect to $q_{i,t}$. Hence (H1) and (H2) imply that π_i is concave. On the other hand, $A_{i,t}(R_{i,t}, K_{i,t})$ is clearly convex and compact. Consequently, there exists at least one Nash-Cournot equilibrium (see theorem 7.1 in Friedman (1977) for details).

Remark: For the sake of simplicity we considered one final consumption market described by a deterministic demand law. The extension to the case of several markets described by stochastic demand laws is straightforward (see Zaccour (1987)).

2.1 The flow of information

We suppose in general, that every firm is perfectly informed on-line so to say, about its own current level of reserves and capacity as well as those of every adversary. At period t, this information is embodied in the prevailing state of the game, namely (R_t, K_t) with $R_t = (R_{i,t})_{i \in I}$ and $K_t = (K_{i,t})_{i \in I}$. Then, at that instant, the problem of firm i is to decide on production $\hat{q}_{i,t}$ and investment $\hat{I}_{i,t}$ as strategies, e.g. as functions of (R_t, K_t) such that the present value $V_{i,t}(R_t, K_t)$ during the periods to go is maximal. The latter quantity satisfies (in equilibrium) the functional equation of dynamic programming

$$V_{i,t}(R_t, K_t) = \max_{q_{i,t}, I_{i,t}} \left[\Pi_{i,t}(q_t, I_t) + \beta_i V_{i,t+1}(R_{t+1}, K_{t+1}) \right]$$

where $q_{j,t}$, $I_{j,t}$ for $j \neq i$ coincide with their equilibrium values $\hat{q}_{i,t}(R_t, K_t)$, $\hat{I}_{j,t}(R_t, K_t)$. We note that the maximizing variables $\hat{q}_{i,t}$, $\hat{I}_{i,t}$ on the right hand side of the above equation depend evidently on (R_t, K_t).

The important and simple point we want to emphasize now is that if $(R_{j,t}, K_{j,t})_{t=o, \ldots, T}$ is specified exogeneously for every agent $j \neq i_o$ (i_o designated), then strategies at any period t depend <u>only</u> on $(R_{i_o,t}, K_{i_o,t})$. In particular this is so if $R_{j,o} = +\infty$ for $j \neq i_o$, and the trajectories $(K_{j,t})_{t=o, \ldots, T}$, $j \neq i_o$, are specified at the outset.

Under such a hypothesis, the feedback equilibrium strategies will appear as completely specified trajectories $(q^*_{i,t}, I^*_{i,t})_{t=o, \ldots, T}$ for all $i \in I$. To see this, let recursively $q^*_{i,t} = \hat{q}_{i,t}(R_{i_o,t}, K_{i_o,t})$ with $(R_{i_o,t}, K_{i_o,t})$ updated in concordance with the prescription of $\hat{q}_{i_o,t}$ and $\hat{I}_{i_o,t}$. Thus equilibrium feedback strategies coincide with their open-loop counterparts.

3. Application to the European gas market

The European gas market can be represented by the network shown in Figure 1.

As one can notice, four types of agents appear in this network. A small number of <u>producers</u>, (namely four Algeria, Holland, Norway and USSR), the <u>shippers</u>, the transmission companies (<u>distributors</u>) who sell gas to the consumers aggregated in consuming regions. The first particularity of this market is the high degree of concentration both at the production and distribution levels. Indeed, the four producers mentioned above were responsible for 98.4% of the exportations towards EEC countries in 1984. On the other hand, the distributors are usually organized as governmentally protected monopolies: they are monopolies with respect to natural gas towards final consumers and do have considerable monopsony power towards producers (Roland et al (1984)). The second important fact is that producers and distributors conclude contracts for a long time period. The main objective of these contracts is to secure heavy investments made by producers in order to develop gas fields.

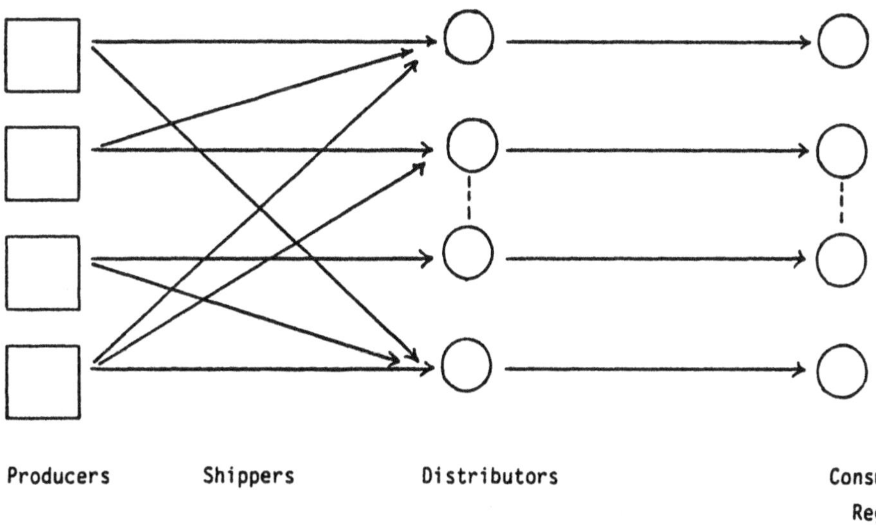

Figure 1: Schematic representation of the European gas market

Recently, the European gas market has been the object of several studies (see for instance Russell (1983), Boucher and Smeers (1984) and Stern (1984)). In particular, two papers by Roland et al (1986) and Haurie et al (1987b) adopted a game theoretical framework to analyze the long term evolution of this market. In both cases, the shippers are represented by transportation costs and hence do not appear as active players in their models. Further, for reasons discussed by Roland et al and adopted by Haurie et al, the distributors are considered dummy players whose only role is the transmission of demand laws to the producers. The problem is then reduced to an oligopoly selling an homogeneous non-renewable resources in a set of competitive markets. Consequently, it seems "natural" to adopt the Nash-Cournot equilibrium as a solution concept. The three main differences between the two models are:

(1) Whereas Roland et al considered deterministic demand laws, Haurie et al adopted stochastic ones. Clearly the latter approach adds some realism to the model.

(2) Roland et al computed Nash equilibria (as well as competitive and monopoly ones) for two static games at two different periods (1983 and 2000). The argument advanced by these authors for not using an intertemporal model a la Hotelling is that high profits are not due to the scarcity of the resources but to their uneven localisation. Moreover, Roland et al noticed that the ratio reserves/production 1983 is almost 70 years in the case of Norway and USSR and more than 100 years for Algerian gas. This argument is no more accepted if one considers a more distant horizon, which is the case of the model of Haurie et al, since the reserves on-shore located in Holland (Groningen) are "scarce". Moreover, according to the results obtained by these authors, these reserves will be depleted at the terminal date (2019) of their model.

3) Finally, whereas Roland et al considered the investment schedules, and hence the production capacities, exogeneous, Haurie et al assumed the investment to be endogeneous in the case of Norway and USSR. They fixed the investment at zero for Algeria (which has a high spare capacity) and Holland (Groningen is fully developed). Notice that the Netherlands hold undeveloped off-shore reserves. The difference in unit costs between the off-shore fields and the Groningen field is as high as 10 to one (Roland et al). Therefore, it does not seem probable that Holland develops these fields in the medium term.

Our point of view is that by involving the hypothesis made in these two studies, one can argue that the European gas market can be fairly approximated by the conditions stated in the previous section. Consequently Nash equilibria in open-loop and feedback strategies coincide. We now look into details:

(1) The objective of maximization of the sum of discounted profits seems acceptable in a long-term perspective. It remains then to choose adequately the initial and terminal dates of the non-cooperative game. Assume that the initial date is the current year (any neighbouring date has no "technical" incident on the remaining). In a finite time horizon model, the terminal date T can be chosen in two manners:

(i) Stipulate T such that the discounted profit of period T + 1 will (almost) be zero i.e. approximate an infinite time horizon model, where the problem of assigning salvage values for the remaining reserves and production capacities is avoided, by a finite time horizon model. If the annual interest rate is in the range 7% - 10%, it suffices with $T \approx 40$ years to have a good approximation.

(ii) If one wishes to consider a less far horizon, then one should add a bequest function of the remaining stocks to the payoff function of each player.

We adopt (i) since it is very common to consider such an horizon in energy studies.

(2) In this case, for the reasons given before, one can consider that the reserves of three of the four-player game (namely Algeria, Norway and USSR) are infinite. Hence, only Holland would be constrained by a state equation governing its reserves depletion.

(3) To end the "proof", it remains necessary to show that the expansion of the production capacities of these three players can be considered exogeneous (or alternatively infinite capacities). For this, one refers to the approach adopted by Roland et al. These authors give two marginal production cost curves (for 1983 and 2000) for each producer. In the short run, the only relevant cost is the variable one, since production capacity is fixed. In the long run, the fixed cost must also be taken into account when computing the profit of each player. For this, Roland et al interpret the marginal production cost curve of year 2000 as an ex-ante curve in the sense that it describes potential production capacity. The marginal production cost function is, in our notations, as follows:

$$C_i(q_{i,t}) = \alpha_i (K_{i,t} - q_{i,t})^{\beta_i} \quad \alpha_i > 0, \beta_i < 0, q_{i,t}, K_{i,t} \geq 0$$

where $K_{i,t}$ is prespecified for 1983 and 2000. The justification to take $K_{i,2000}$ as constant is that the decision to expand the actual production capacity has been already taken and is, in some sense, irreversible (for example developing the Norwegian Troll field whose deliveries will start only around year 2000 has already been taken and even developing the field has started). Finally, to meet the conditions defined in the previous section, one has to make the assumption that:

$$K_{i,t} = K_{i,1983} \qquad t = 1983, \ldots\ldots, 1999$$

$$K_{i,t} = K_{i,2000} \qquad t = 2000, \ldots\ldots, T$$

Whereas the first part of this hypothesis seems very reasonable due to investment delays, the second part is undoubtedly strong and should be refined.

If our reasoning summarized in (1) - (3) is accepted then Nash equilibria in open-loop and feedback strategies coincide. This clearly means that we could easily compute for this market credible (subgame perfect) equilibria, The next and final section reviews recent advances in the computation of Nash equilibria.

4. Conclusion

Computation of Nash-Cournot equilibria has been an active area of research in recent years. Murphy et al (1982) suggested an algorithm based on the replacemnet of the original equilibrium problem by a sequence of constrained concave maximization problems whose solutions converge towards the solution of the original problem. On the other hand, Gabay and Moulin (1980) showed that under some conditions, essentially monotonicity of marginal revenue, the conditions for a Nash-Cournot equilibrium are equivalent to a variational inequality which can be solved by several well-known algorithms (see Pand and Chan (1982) and Dafermos (1984)). Harker (1984), Marcotte (1984), Cohen and Chaplais (1986) and Flam and Ben-Israel (1987) developed this approach in the case of static Nash equilibrium. Harker (1986), Marcotte (1985) and Nagurney (1986) extended it to the case where the imperfect competition is defined on a network. Finally, Zaccour (1987) adapted this variational inequality approach to the case where each player controls a set of production units, dynamic and several markets described by stochastic demand laws.

BIBLIOGRAPHY

Boucher, J. Smeers, Y.(1984) Simulation of the European Gas Market up to the year 2000. CORE discussion paper, No 8448.

Cohen, G., Chaplais, F. (1986) Algorithmes numériques pour les équilibres de Nash. R.A.I.R.O. APPII, vol 20, pp. 273-293.

Dafermos, S.C. (1983) An Iterative Scheme for Variational Inequalities. Mathematical Programming, vol 29, pp 40-47.

Esvaran, M., Levis, T. (1985) Exhaustible Resources and Alternative Equilibrium Concepts. Canadian Journal of Economics, XVIII, No 3, August, P8.459-473.

Flåm, S.D., Ben-Israel, A. (1987) A continuous Approach to Oligopolistic Market Equilibrium, Working Paper No 387, Department of Economics, University of Bergen, Norway.

Friedman, J.W. (1977) Oligopoly and the Theory of Games. North Holland.

Gabay, D., Moulin, H. (1980) On the Uniqueness and Stability of Nash-Equilibria in Noncooperative Games. In A. Bensoussan, P. Kleidorfer and C.S. Tafiero, eds., Applied Stochastic control in Econometrics and Management Science, North Holland.

Harker, P.T. (1984) A Variational Inequality Approach for the Determination of Oligopolistic Market Equilibrium. Mathematical Programming, vol. 30, pp. 105-111.

Harker, P.T. (1986) Alternative Models of Spatial Competition. Operations Research, vol. 34, No 3, pp. 410-425.

Haurie, A., Zaccour, G., Legrand, J., Smeers, Y. (1987a). A Dynamic Stochastic Nash-Cournot Model for the European Gas Market. Cahier du GERAD, G.87-24, submitted to Operations Research.

Haurie, A., Zaccour, G., Smeers, Y. (1987b) Stochastic Equilibrium programming for Dynamic Oligopolistic Markets. Cahier du GERAD, 87-33, to appear in Journal of Optimization Theory and Applications.

Marcotte, P. (1984) Quelques notes et résultats nouveaux sur le problème d'équilibre d'un oligopole. R.A.I.R.O. RECHERCH OPERATIONNELLE, vol. 18, No 2, pp. 147-171.

Marcotte, P. (1985) Algorithms for the Network Oligopoly Problem. Centre de recherche sur les transports, Publication No 426.

Murphy, F.H., Sherali, H.D., Soyster, A.L. (1982) A Mathematical Programmming Approach for Determining Oligopolistic Market Equilibria. Mathematical Programming, vol. 24, pp 92-106.

Nagurney, A. (1986). Competitive Equilibrium Problems, Variational Inequalities and Regional Science. Journal of Regional Science, vol. 27, No 4, pp. 503-517.

Newberry, D.M. (1984) The Economics of Oil. In Mathematical Methods in Economics, edited by F. Van Den Ploeg, John Wiley & Sons, Ltd

Pang, J.S., Chan, D. (1982) Iterative Methods for Variational and Complementarity Problems, Mathematical Programming, vol 24, pp 284-313.

Reinganum, J.F., Stockey, N.L. (1985) Oligopoly Extraction of a common Property Natural Resource : the Importance of the Period of commitment in Dynamic Games. International Economic Review, vol 26, No 1, pp 161-173

Roland, K., Thonstad, K., Mathiesen, L. (1986) The European Natural Gas Market: Degrees of Monopoly Power. Paper presented at the conference on natural gas contracts and markets, Geilo, Jan. 6-9.

Russel, J. (1983) Geopolitics of Natural Gas. Ballinger Publishing Company.

Salant, S.W. (1982) Imperfect Competition in the International Energy Market: a Computerized Nash-Cournot Model. Operations Research, vol. 30, No 2, pp 252-280.

Selten, R. (1975) Reexamination of the Perfectness concept for Equilibrium Points in Extensive Games. International Journal of Games Theory, vol. 4, pp 25-55.

Stern, J.P. (1984) International gas trade in Europe : the Policies of Exporting and Importing Countries. Heineman Educational Books, London.

Zaccour, G. (1987) Théorie des jeux et marchés énergétiques : marché européen du gas naturel et échanges d'électricité, Cahier du GERAD, G-87-34.

TIME–DEPENDENT COOPERATION IN GAMES

L. Gao[1], A. Jakubowski[2], M.B. Klompstra, G.J. Olsder
Department of Technical Mathematics and Informatics
Delft University of Technology
P.O.Box 356, 2600 AJ Delft, The Netherlands

Abstract
The behaviour of players in a dynamic game is considered if they are entitled to –and agree upon– cooperation. This cooperation will depend on time, and during time intervals in which there is no cooperation, the Nash–mode of play is assumed. Cooperation occurs if all players benefit by it.

1. Introduction

The theory of both cooperative and noncooperative dynamic games has been well developed [1], [3]. In reality, however, one encounters situations in which changes take place from cooperation to noncooperation and vice versa. Another phenomenon that happens in practice is the formation of coalitions between a subset of players among all the players involved in a game. Moreover such coalitions may change in time [5]. The theory of time–dependent cooperation and time–dependent coalitions is still in its infancy. This paper aims at a better understanding of time–dependent cooperation.

We will study the behaviour of players in a dynamic game if they are entitled to (and agree upon) cooperation. This cooperation will depend on time and during time intervals in which there is no cooperation, the Nash–mode of play is assumed. There are no binding contracts and cooperation occurs if all players benefit from it. If at least one player does not favour cooperation, the evolution of the game will be according to the Nash equilibrium concept. In the latter part of the paper we also consider the situation in which the decision of cooperation versus noncooperation is influenced stochastically. We are interested in the switching moments from the cooperative to the noncooperative mode of play and vice versa. For the infinite horizon game, the

[1]visitor from Northeast University of Technology, Shenyang, China

[2]visitor from Polish Academy of Sciences, Warsaw, Poland

problem has not yet been solved in its most general setting, in which one needs the dynamic programming equations including an arbitrary number of future switching moments. Instead, the decision of the players whether to cooperate or not to cooperate, assumes no future switching moments. Since this decision is made continuously in time, in reality many such switches may (and will) occur.

The emphasis will be on linear–quadratic games. The evolution of the game is considered both deterministic and stochastic. This latter case refers to system noise and in the infinite horizon game a discount factor has been added such as to make the outcome finite. Full state information is assumed and the players also know each other's controls up to the current time. Therefore cheating is excluded and will not be considered. If only state information were available, the situation would be more subtle [6], [9].

The evolution of the game very much depends on the definition of cooperation, or, to say it differently, on which Pareto–solution is agreed upon among the players. It will be shown that for a (reasonably looking) definition only cooperation occurs and therefore no switches to the noncooperative mode exist. For other definitions such switches do occur, except for the case that the dimension n of the state vector x equals one. Numerical examples will be given. Existence of solutions in these games has been considered. If the decision whether to cooperate or not to cooperate has a stochastic component, it depends on the so–called cooperative probability functions, how often the switching moments occur.

2. Problem Statement

With the exception of sections 7 and 8 the problems to be considered in this paper belong to the class of two–person nonzero–sum infinite horizon differential games. The emphasis will be on linear dynamic and quadratic cost functions (in particular in sections 4, 5 and 6).

The general state equation is

$$\dot{x}(t) = f(x(t), u_1(t), u_2(t), t) \quad , t \in [0,\infty) , \tag{2.1}$$

where $x(t) \in \mathbb{R}^n$, $u_i(t) \in \mathbb{R}^{m_i}$ The vector x is called the state, the vectors u_1 and u_2 are the controls to be chosen by players P_1 and P_2 respectively. Each player has a (scalar) cost function

$$J_i(u_1, u_2) = \int_0^\infty h_i(x(t), u_1(t), u_2(t), t)dt \quad , i = 1, 2 , \tag{2.2}$$

which he wants to minimize in a way to be determined later on. It is assumed that both players know the state equation as well as the cost functions. During the evolution of the game it is assumed that both players know the state (full state information) and each other's control functions up to the current time. Sometimes we will write $u_i(t) = \gamma_i(x(t))$, which refers to a feedback control law; the strategy γ_i maps the state (information) into the control (the action). When speaking about strategies, it will always be assumed that this mapping is meant. Sometimes a restriction is made to a specific class of strategies; $\gamma_i \in \Gamma_i$, where Γ_i is the admissible strategy space.

An admissible strategy pair (γ_1^*,γ_2^*) constitutes a *Nash equilibrium solution* if the following inequalities are satisfied:

$$J_1(\gamma_1^*,\gamma_2^*) \le J_1(\gamma_1,\gamma_2^*) \; ; \; J_2(\gamma_1^*,\gamma_2^*) \le J_2(\gamma_1^*,\gamma_2) \; , \tag{2.3}$$

for all $\gamma_i \in \Gamma_i$. If the players choose their strategies according to this solution concept, they are said to play *noncooperatively*. The *value functions*, i.e. the minimum cost to go from any initial state x and any initial time $\tau \in [0,\infty)$, are given by

$$V_i^N(x,\tau) = \min_{\gamma_i(\cdot)} \int_\tau^\infty h_i(x, \gamma_1, \gamma_2, t)dt \quad , i = 1, 2 \tag{2.4}$$

subject to (2.1) and initial condition $x(\tau) = x$. In (2.4) it is assumed the argument γ_j of the h_i–function with $j \neq i$ is held fixed. The superscript N in (2.4) has been added such as to indicate the underlying Nash equilibrium.

Another solution concept is the one named after Pareto. An admissible strategy pair (γ_1^*,γ_2^*) is called *Pareto optimal* if

$$J_i(\gamma_1,\gamma_2) \le J_i(\gamma_1^*,\gamma_2^*) \quad , i = 1, 2 \; ,$$

where at least one of the inequalities is strict, does not admit a solution (γ_1,γ_2). In other words, there is no joint solution (γ_1,γ_2) which is better for both players than the solution (γ_1^*,γ_2^*). If certain convexity conditions are satisfied [4], one can show that Pareto–solutions can be obtained by

$$\min_{u_1,u_2} J \; ; \; J = \mu_1 J_1 + \mu_2 J_2 \; , \tag{2.5}$$

the scalar coefficients μ_1 and μ_2 satisfy $\mu_i > 0$ and $\mu_1 + \mu_2 = 1$. For different choices of μ_1,μ_2 one finds different Pareto solutions. If the players choose their strategies according to the Pareto optimality concept, they are said to play *cooperatively*. The minimal costs to go which those players face when they play cooperatively are indicated by $V_i^C(x,\tau)$, $i = 1, 2$, where τ stands for the current time and x for the current state. The superscript C for the V_i (value function) indicates that the costs refer to cooperative play. In the next section the V_i^C–functions will be discussed in more detail.

In this paper the players have an option to play cooperatively or noncooperatively. If at least one of them favours the noncooperative mode of play, the strategies for both players will be determined according to the Nash equilibrium. If both players favour cooperation, and this is the case when

$$V_i^C(x,\tau) \le V_i^N(x,\tau) \quad , i = 1, 2 \; , \tag{2.6}$$

then the strategies will be determined according to the Pareto concept. If the inequalities (2.6) are not fulfilled, the players will play noncooperatively. Since (2.6) depends on x as well as τ, switches from the cooperative mode to the noncooperative mode, and vice versa, seem to be possible. We will investigate whether and when such switches occur.

The decision whether to cooperate or not depends on the inequalities (2.6). The functions V_i^C and V_i^N do not take possible future switches into account, since for the calculation of V_i^N (and also of V_i^C) it is assumed that the mode of play remains fixed for the whole future.

3. Modes of cooperation

As already mentioned, The Pareto solution is not unique; different μ_i–values lead to different solutions. We will consider two different solutions.

Cooperative solution I. If the current state and time are x and τ respectively, then the factors μ_i in (2.5) are chosen as

$$\mu_i = V_i^N(x,\tau)/(V_1^N(x,\tau) + V_2^N(x,\tau)) \,, \tag{3.1}$$

where the tacit assumption has been made that $V_i^N(x,\tau) > 0$, $i = 1, 2$. The optimal strategies for this cooperative solution are found from (2.5). The individual costs of the players are obtained by substituting these optimal strategies in their respective cost functions.

Cooperative solution II. The starting point for this solution is $\mu_1 = \mu_2 = \frac{1}{2}$. This leads to the minimization of the joint costs $J_1 + J_2$. Each player now takes a share of these costs. If we define $J^* = \min\,(J_1 + J_2)$, then the costs for player P_i are $\lambda_i \cdot J^*$, where λ_i is the same as μ_i as defined in (3.1). The minimizing controls will be indicated by u_1^*, u_2^*.

In both cooperative solution concepts the costs by the players are indicated by $V_i^C(x,\tau)$. Note that in cooperative solution I each player faces his own costs, whereas with cooperative solution II first the joint costs are minimized and then a split up between the players is defined (there will be side payments in the latter case).

Theorem. If the players have the options to cooperate and not to cooperate, and for cooperation the solution II is chosen, then only cooperation takes place. There will be no switches to the noncooperative mode.

Proof. According to the definition of u_i^*,

$$J^* = J(u_1^*, u_2^*) \le J_1(u_1, u_2) + J_2(u_1, u_2)$$

for arbitrary u_1, u_2 and therefore also for the controls u_1^N, u_2^N which constitute the Nash equilibrium. Hence $J^* \le V_1^N + V_2^N$, and

$$V_i^C = \mu_i\, J^* = \frac{V_i^N}{V_1^N + V_2^N} \cdot J^* \le V_i^N \quad, i = 1, 2 \;,$$

which means that it is always advantageous for the players to cooperate. □

With respect to the cooperative solution I, switches from cooperative to noncooperative mode and vice versa do exist as will be seen in section 5.

4. Linear–quadratic games.

In this section it is assumed that

$$f(x, u_1, u_2, t) = Ax + B_1u_1 + B_2u_2,$$
$$h_i(x, u_1, u_2, t) = \tfrac{1}{2}(x'Q_ix + u_i'R_{ii}u_i + u_j'R_{ij}u_j), \qquad (4.1)$$
$$i, j = 1, 2 ; j \neq i.$$

The symbol $'$ denotes transpose. The constant matrices A, B_1, B_2, Q_i, R_{ii} and R_{ij} ($i \neq j$) have appropriate sizes. It is assumed that $Q_i = Q_i'$, $R_{ii} = R_{ii}' > 0$, $R_{ij} = R_{ij}'$. It is well known [1] that the optimal feedback strategies for the Nash equilibrium concept are given by (provided they exist):

$$u_i^N = \gamma_i^N(x) = -R_{ii}^{-1}B_i'M_ix, \; i = 1, 2 \qquad (4.2)$$

where the $n*n$ matrices M_i satisfy

$$0 = M_iA + A'M_i + Q_i - M_iB_iR_{ii}^{-1}B_i'M_i - M_iB_jR_{jj}^{-1}B_j'M_j +$$
$$-M_jB_jR_{jj}^{-1}B_j'M_i + M_jB_jR_{jj}^{-1}R_{ij}R_{jj}^{-1}B_j'M_j ; \qquad (4.3)$$
$$i, j = 1, 2; i \neq j.$$

It is also known that [7] if

i) solutions M_1 and M_2 to (4.3) exist

ii) $A_N \triangleq A - B_1R_{11}^{-1}B_1'M_1 - B_2R_{22}^{-1}B_2'M_2$ is asymptotically stable (4.4)

iii) $Q_i + M_jB_jR_{jj}^{-1}R_{ij}R_{jj}^{-1}B_j'M_j \geq 0$, $i \neq j$,

then the strategies (4.2) exist, V_1^N and V_2^N are finite and are given by

$$V_i^N(x(\tau),\tau) = \tfrac{1}{2}x'(\tau)M_ix(\tau). \qquad (4.5)$$

In [8] some weaker conditions are given for the case that $B_1 = B_2$. The costs of (4.5) are to be compared with the costs obtained by cooperative play. Towards that end consider

$$J = \mu_1J_1 + \mu_2J_2 = \tfrac{1}{2}\int_0^{\infty}(x'Qx + u_1'R_1u_1 + u_2'R_2u_2)dt, \qquad (4.6)$$

where $Q = \mu_1Q_1 + \mu_2Q_2$; $R_i = \mu_iR_{ii} + \mu_jR_{ji}$ ($i \neq j$). The minimizing feedback strategies of (4.6) are

$$u_i^C = \gamma_i^C(x) = -R_i^{-1}B_i'M_0x, \; i = 1, 2 \qquad (4.7)$$

where the $n*n$ matrix M_0 satisfies

$$0 = M_0A + A'M_0 + Q - M_0(B_1R_1^{-1}B_1' + B_2R_2^{-1}B_2')M_0 . \qquad (4.8)$$

These strategies exist if $Q \geq 0$, $R_i > 0$, μ_i is constant, the matrix pair $(A,[B_1\ B_2])$ is controllable and if the matrix pair (A,C) is observable, where $Q = C'C$. The minimal joint costs are given by

$$J^* = \frac{1}{2} x'(\tau)M_0x(\tau) . \qquad (4.9)$$

The players now decide on the cooperative solution I and the μ_i-values of (3.1) are substituted into (4.6). Some analysis shows that the individual costs are

$$V_i^C(x(\tau),\tau) = \frac{1}{2} x'(\tau)P_ix(\tau) , i = 1, 2 , \qquad (4.10)$$

where the $n*n$ matrix $P_i \geq 0$ is the solution to the Lyapunov equation

$$P_iA_C + A_C'P_i + Q_i^C = 0 , \qquad (4.11)$$

in which

$$\left.\begin{aligned} &A_C = A - B_1K_1 - B_2K_2 , \\ &Q_i^C = Q_i + K_i'R_{ii}K_i + K_j'R_{ij}K_j \ (i \neq j) , \\ &K_i = R_i^{-1}B_i'M_0 \end{aligned}\right\} \qquad (4.12)$$

It is worthwhile to note, however, that the coefficients μ_i as given by (3.1) are functions of $x(\tau)$! Therefore the costs as given by (4.9) and (4.10) are not purely quadratic in the current state because the matrices M_0 and P_i depend on $x(\tau)$ as well! Also note that during noncooperative play the game evolves according to $\dot{x} = A_N x$ and during cooperative play according to $\dot{x} = A_C x$. In principle μ_i and A_C change continuously with time through the dependence on $x(\tau)$. We will, however, consider the question whether to cooperate or not, not continuously, but at time steps t_i, $i = 0, 1, 2, \ldots$, with $t_{i+1} = t_i + \Delta t$, $\Delta t > 0$. Between these time steps μ_i, and therefore also A_C, will not be adjusted.

5. Examples

Example 1. Consider the following scalar dynamics and cost functions:

$$\dot{x} = -x + u_1 + u_2 , x(0) = x_0 , t \in [0,\infty) ,$$

$$J_1 = \frac{1}{2} \int_0^\infty (20x^2 + u_1^2)dt ,$$

$$J_2 = \frac{1}{2} \int_0^\infty (27x^2 + u_2^2)dt .$$

The positive M_i-solutions of (4.3) are $M_1 = 2$ and $M_2 = 3$. This means that the minimal costs for the players in the Nash sense are x_0^2 and $\frac{3}{2} x_0^2$ respectively. According to (3.1), $\mu_1 = .4$ and $\mu_2 = .6$. With these values of μ_i the scalars P_1 and P_2 as defined by (4.11) become $P_1 = 2.4652$, $P_2 = 1.9930$. It now follows that $M_1 < P_1$ and $M_2 > P_2$ which means that P_2 wants to cooperate but that P_1 prefers noncooperation. The ultimate decision therefore is no cooperation at all and this is valid for all $t \in [0,\infty)$. □

Example 2. The following example also has scalar dynamics:

$$\dot{x} = -5x + \frac{1}{2}u_1 + \frac{1}{2}u_2 \,,\; x(0) = x_0 \,,\; t \in [0,\infty) \,,$$

$$J_1 = \frac{1}{2}\int_0^{\infty}(23x^2 + u_1^2)dt \,,$$

$$J_2 = \frac{1}{2}\int_0^{\infty}(23x^2 + u_2^2)dt \,.$$

The equations for M_1, M_2, M_0, P_1 and P_2 are respectively

$$-10M_i + 23 - \frac{1}{2}M_i^2 - \frac{1}{4}M_1M_2 = 0 \,,\; i = 1, 2 \,,$$

$$(4\mu_1 + 4\mu_2)^{-1}M_0^2 + 10M_0 - 23 = 0 \,,$$

$$2(5 + (4\mu_1 + 4\mu_2)^{-1}M_0)P_i = 23 + \{(2\mu_i)^{-1}M_0\}^2 \,,\; i = 1, 2 \,.$$

The (positive) solutions are $M_1 = M_2 = 2$, therefore $\mu_1 = \mu_2 = \frac{1}{2}$, and $M_0 = 1.9282$, $P_1 = P_2 = 1.9282$. It follows that $P_i < M_i$, $i = 1, 2$ and hence the players decide to cooperate and this cooperation will last forever. □

Example 3. The dynamics and costfunctions are

$$\dot{x} = \begin{bmatrix}-1 & 2\\ 1 & -3\end{bmatrix}x + \begin{bmatrix}0\\1\end{bmatrix}u_1 + \begin{bmatrix}1\\0\end{bmatrix}u_2 \,,\; x(0) = x_0 = \begin{bmatrix}-2\\5\end{bmatrix} \,,\; t \in [0,\infty) \,,$$

$$J_1 = \frac{1}{2}\int_0^{\infty}(x'\begin{bmatrix}9 & -11\\ -11 & 18\end{bmatrix}x + u_1^2)dt \,,$$

$$J_2 = \frac{1}{2}\int_0^{\infty}(x'\begin{bmatrix}4 & -4\\ -4 & 37\end{bmatrix}x + u_2^2)dt \,.$$

The matrices M_1 and M_2 are the positive definite solutions of (4.3);

$$M_1 = \begin{bmatrix}1 & -1\\ -1 & 2\end{bmatrix} \,,\; M_2 = \begin{bmatrix}2 & 1\\ 1 & 4\end{bmatrix} \,,$$

and hence $V_1^N(x_0,0) = 37$, $V_2^N(x_0,0) = 44$. This leads to $\mu_1 = \mu_1(x_0) = 0.4568$, $\mu_2 = \mu_2(x_0) = 0.5432$. For the solutions of M_0, P_1 and P_2, all depending on x_0, we find

$$M_0 = \begin{bmatrix}1.32 & -0.18\\ -0.18 & 2.43\end{bmatrix} \,,\; P_1 = \begin{bmatrix}1.02 & -0.46\\ -0.46 & 2.71\end{bmatrix} \,,\; P_2 = \begin{bmatrix}1.60 & 0.23\\ 0.23 & 2.27\end{bmatrix} \,.$$

Now

$$V_1^C(x_0,0) = 40.45 > V_1^N(x_0,0) \;;\; V_2^C(x_0,0) = 29.24 < V_2^N(x_0,0)$$

which means that player P_2 is in favour of cooperation and P_1 is not, at least during the initial phase of the game. The joint decision is no cooperation, based on inequalities which assume no cooperation for the whole future! However, every $\Delta t = 0.1$ units of time the players will compare V_i^C and V_i^N again. It is easily seen that μ_1 and μ_2 will change with respect to time. The result is given in the following table, where (N)C stands for in favour of (no)cooperation.

t	P_1	P_2	joint decision
.0	NC	C	NC
.1	NC	C	NC
.2	NC	C	NC
.3	NC	C	NC
.4	NC	C	NC
.5	C	C	C
.6	C	C	C
.7	C	C	C
.8	C	C	C
.9	C	NC	NC
1.0	C	NC	NC
1.1	C	C	C
1.2	NC	C	NC
1.3	NC	C	NC
1.4	NC	C	NC
1.5	NC	C	NC
>1.5	NC	C	NC

It turns out in this example that M_0, P_1 and P_2 converge to constant matrices (after $t = 3$ there are no noticeable changes any more);

$$\overline{M}_0 = \begin{bmatrix} 1.29 & -0.02 \\ -0.02 & 1.58 \end{bmatrix}, \ \overline{P}_1 = \begin{bmatrix} 1.95 & 0.10 \\ 0.10 & 7.30 \end{bmatrix}, \ \overline{P}_2 = \begin{bmatrix} 1.22 & -0.03 \\ -0.03 & 0.96 \end{bmatrix}. \qquad \square$$

6. Stochastic extension

In this section the stochastic extension is considered of the linear quadratic infinite horizon game. The dynamics now are

$$\dot{x} = Ax + B_1u_1 + B_2u_2 + G\xi \ , \ x(0) = x_0 \ , \ t \in [0,\infty) \ . \tag{6.1}$$

Except for G and ξ the other quantities are as before. The vector ξ has m components and $\xi(t)$ is a Gaussian white noise process [2] with average zero and covariance matrix Σ. Matrix G is constant and has size n*m. The cost functions are

$$J_i(u_1,u_2) = E[\tfrac{1}{2}\int_0^{\infty} e^{-\beta t}(x'Q_ix + u_i'R_{ii}u_i + u_j'R_{ij}u_j)dt] \ , \tag{6.2}$$

$$i, j = 1, 2 \ ; \ i \neq j \ ,$$

where $\beta > 0$ is a discount factor. The operator E stands for expectation.

Noncooperative feedback solution. This solution [1] has again the form

$$u_i^N = \gamma_i(x) = -R_{ii}^{-1}B_i'M_ix \, , \, i = 1, 2 \, ,$$

where the matrices M_i satisfy equations (4.3) with the exception that the matrix A in these equations is now replaced by $A - \frac{1}{2}\beta I$, I denoting the unit matrix. The value functions, i.e., the outcome of the cost functions (if initial time and state are τ and $x(\tau)$) in which the optimal strategies are substituted, are

$$W_i^N(x(\tau),\tau) = \frac{1}{2}e^{-\beta\tau}x'(\tau)M_ix(\tau) + \frac{1}{2\beta}e^{-\beta\tau}\,\text{trace}\,(G'M_iG), \, i = 1, 2 \, .$$

Cooperative feedback solution. The criterion is $J = \mu_1J_1 + \mu_2J_2$, with $\mu_i = W_i^N/(W_1^N + W_2^N)$. Similarly to the deterministic case, we write

$$J(u_1,u_2) = E[\frac{1}{2}\int_0^\infty e^{-\beta t}(x'Qx + u_1'R_1u_1 + u_2'R_2u_2)dt] \triangleq E[\int_0^\infty h(x_1, u_1, u_2, t)dt],$$

with Q and R_i as before. In order to find the optimal strategies, the stochastic version of the Hamilton–Jacobi–Bellman equation [2] can be used:

$$\frac{\partial W}{\partial t} + \frac{1}{2}\,\text{trace}\,(G'\frac{\partial^2 W}{\partial x^2}G) + \min_{u_1,u_2}(\frac{\partial W}{\partial x}f + h) = 0 \, ,$$

where $W(x(\tau),\tau)$ is the value fuction if the initial time and state are τ and $x(\tau)$ and where $f = Ax + B_1u_1 + B_2u_2$. The optimal strategies are $u_i^* = -R_i^{-1}B_i'M_0x$, where the n*n matrix M_0 satisfies (4.8) if A is replaced by $A - \frac{1}{2}\beta I$. The value functions $W_i^C(x(\tau),\tau)$ for the individual players in case of cooperation can now also be found. For that reason the optimal strategies must be substituted into (6.1) and (6.2). Subsequent analysis leads to

$$W_i^C(x(\tau),\tau) = \frac{1}{2}e^{-\beta\tau}x'(\tau)P_i^Cx(\tau) + \frac{1}{2\beta}e^{-\beta\tau}\,\text{trace}\,(C_iP_i^RC_i') \, , \tag{6.3}$$

where P_i^C and P_i^R are the positive definite solutions of

$$(A_C - \frac{1}{2}\beta I)'P_i^C + P_i^C(A_C - \frac{1}{2}\beta I) + Q_i^C = 0 \, ,$$

$$(A_C - \frac{1}{2}\beta I)P_i^R + P_i^R(A_C - \frac{1}{2}\beta I)' + Q_\Sigma = 0 \, ,$$

where $Q_\Sigma = G\Sigma G'$ and where A_C, Q_i^C are the same quantities as defined in (4.12). The quantity C_i in (6.3) is defined through $Q_i^C = C_i'C_i$.

7. Cooperative probabilities

In this section we confine ourselves to two–person finite horizon deterministic difference games. The number of stages is K. The state equation and cost functions are, respectively,

$$\left.\begin{aligned} &x(k+1) = f(x(k), u_1(k), u_2(k)) \, , \, k = 0, 1,\ldots, K \, , \\ &J_i = \sum_{k=0}^{K} h_i(x(k), u_1(k), u_2(k)) \, , \, i = 1, 2 \, . \end{aligned}\right\} \tag{7.1}$$

As before, $x \in \mathbb{R}^n$, $u_i \in \mathbb{R}^{m_i}$; it is assumed that during the evolution of the game the players have full state information. If cooperation takes place, the players jointly minimize $J_1 + J_2$ (or, properly speaking, the part of $J_1 + J_2$ from the current time till the final time) and of these costs P_i gets his share $\alpha_i(J_1 + J_2)$, where α_i, called the distribution factors, are given positive numbers which add up to one. These distribution factors are assumed to be constant; they do not change with k. The players themselves must decide about their shares of the total costs $J_1 + J_2$, before the actual game starts. The choice of α_i is decided upon via cooperative probability functions to be introduced now.

The cooperative probability functions $\beta_i(\alpha_i)$ express the probability $(0 \le \beta_i \le 1)$ that player P_i would like to cooperate if α_i is given. The ultimate probability whether cooperation takes place or not is given by $\beta \triangleq \beta_1(\alpha_1)\cdot\beta_2(\alpha_2)$. If α_1 is close to zero, then P_1's share of $J_1 + J_2$ is small and hence his willingness to cooperate is large. On the other hand, if α_1 is close to one, this willingness is small. Therefore it is reasonable to assume that the prescribed function $\beta_1(\alpha_1)$ is a nonincreasing function with $\beta_1(0) = 1$, $\beta_1(1) = 0$. Similarly, $\beta_2(\alpha_2)$ has the same properties. Since for a two–person game $\alpha_1 + \alpha_2 = 1$, β_2 can be expressed as a function of α_1 also and with respect to α_1 the function β_2 is nondecreasing. Cooperation/noncooperation is purely determined by the probability β and this holds true at each time step; in particular there is no correlation in these probabilities with respect to k.

Now that the rules of the game have been set, the remainder of this section will be devoted to two aspects: i) what do the value functions look like given the parameter $\alpha \triangleq \alpha_1 = 1 - \alpha_2$, and ii) what is a good choice for α. The underlying assumption is that the functions $\beta_i(\alpha_i)$, $i = 1, 2$, are given and known to the players. If at the current stage k the mode of play is noncooperative, then the cost functions will be denoted by $J_i^N(x(k),k)$. If this mode of play is cooperative, the cost function will be denoted by $J_i^C(x(k),k)$. Denote $V_i^N = \min\limits_{u_i} J_i^N$ and $V_i^C = \min\limits_{u_1,u_2} J_i^C$.

Application of the principle of dynamic programming leads to the following recurrence relations:

$$\left.\begin{aligned}
V_1^N(x(k),k) &= \min_{u_1(k)} \{h_1(x(k), u_1(k), u_2(k)) + \\
&\quad + \beta\alpha V^C(x(k+1),k+1) + (1-\beta)V_1^N(x(k+1),k+1)\}\,, \\
V_2^N(x(k),k) &= \min_{u_2(k)} \{h_2(x(k), u_1(k), u_2(k)) + \\
&\quad + \beta(1-\alpha)V^C(x(k+1),k+1) + (1-\beta)V_2^N(x(k+1),k+1)\}\,, \\
V_i^C(x(k),k) &= \min_{u_1(k),u_2(k)} \alpha_i\{h_1(x(k), u_1(k), u_2(k)) + \\
&\quad + h_2(x(k), u_1(k), u_2(k)) + \beta V^C(x(k+1),k+1) + \\
&\quad + (1-\beta)\sum_{j=1}^{2} V_j^N(x(k+1),k+1)\}\,, \; i = 1, 2\,.
\end{aligned}\right\} \qquad (7.2)$$

If we specialize to the linear–quadratic case, such that the functions f and h_i, i = 1, 2, are as in eq.(4.1), then eqs.(7.2) can be solved explicitly. We try the following substitutions into (7.2):

$$V_i^N(x(k),k) = \frac{1}{2} x'(k)M_i(k)x(k) \; , \; i = 1, 2$$

$$V^C(x(k),k) = \frac{1}{2} x'(k)P(k)x(k) \; ,$$

and it turns out that the M_i– and P– matrices can be solved backwards in time.

$$L_i^N(k) = \{R_{ii} + B_i'N_i(k+1)[I - B_j(R_{jj} + B_j'N_j(k+1)B_j)^{-1}B_j'N_j(k+1)]B_i\}^{-1} \cdot$$
$$\cdot B_i'N_i(k+1)\{B_j(R_{jj} + B_j'N_j(k+1)B_j)^{-1}B_j'N_j(k+1) - I\} \cdot A, \; j \neq i, \; i = 1, 2,$$

$$L_i^C(k) = \{R_i + B_i'N_0(k+1)[I - B_j(R_j + B_j'N_0(k+1)B_j)^{-1}B_j'N_0(k+1)]B_i\}^{-1} \cdot$$
$$\cdot B_i'N_0(k+1)\{B_j(R_j(R_j \; + \; B_j'N_0(k+1)B_j)^{-1}B_j'N_0(k+1) \; - \; I\} \cdot A, \; j \neq i, \; i = 1, 2,$$

$$N_i(k) = \alpha_i\beta P(k) + (1-\beta)M_i(k) \; , \; i = 1, 2,$$

$$N_0(k) = \beta P(k) + (1-\beta)(M_1(k) + M_2(k)) \; ,$$

$$M_i(k) = Q_i + L_i^{N'}R_{ii}L_i^N + L_j^{N'}R_{ij}L_j^N + \alpha_i\beta A_C'P(k+1)A_C + (1-\beta)A_N'M_i(k+1)A_N \; , \; j \neq i, \; i = 1, 2,$$

$$P(k) = Q + L_1^{C'}R_1L_1^C + L_2^{C'}R_2L_2^C + \beta A_C P(k+1)A_C + (1-\beta)A_N(M_1(k+1) + M_2(k+1))A_N \; ,$$

$$A_C = A + B_1L_1^C + B_2L_2^C \; ; \; A_N = A + B_1L_1^N + B_2L_2^N,$$

$$M_i(K) = Q_i(K) \; ; \; P(K) = Q(K) = Q_1(K) + Q_2(K).$$

Whereever in the above formulas no arguments are mentioned in the right hand side, the argument k+1 is meant. Also, it is tacitly assumed that the inverse matrices will exist in these formulas. The optimal strategies can be obtained from the following expressions:

$$u_i^N(k) = -(R_{ii} + B_i'N_i(k+1)B_i)^{-1}B_i'N_i(k+1)(Ax(k) + B_ju_j^N(k)) \; , \; j \neq i, \; i = 1, 2 \; ;$$

$$u_i^C(k) = -(R_i + B_i'N_0(k+1)B_i)^{-1}B_i'N_0(k+1)(Ax(k) + B_ju_j^C(k)) \; , \; j \neq i, \; i = 1, 2.$$

This section will be concluded with some remarks on a proper choice for $\alpha(=\alpha_1=1-\alpha_2)$. A reasonable choice seems to be $\alpha^* = \arg\max_\alpha \beta(\alpha) = \arg\max \beta_1(\alpha)\cdot\beta_2(1-\alpha)$. For player P_i the best choice would be $\alpha_i^* = \arg\min_{\alpha_i} \alpha_i\beta V_i^C(x(0),0) + (1-\beta)V_i^N(x(0),0)$. Of course, not necessarily $\alpha_1^* + \alpha_2^* = 1$ and therefore α_1^* (or α_2^*) can only be enforced if P_1 (or P_2) dominates the game and can himself decide on α_1^*.

8. Exact state–dependent cooperation/noncooperation

In the foregoing sections rules have been given according to which the players will play cooperatively or noncooperatively. In the first sections these decisions were based on the assumption that no new switches from cooperation to noncooperation or vice versa would take place. In the last section (section 7) the decision about cooperation depended on probability functions in which the state did not appear.

In this section a few remarks will be made about the exact solution. The starting point is eq.(7.1) in which f and h_i are as in eq.(4.1), with $R_{ij} = 0$, $i \neq j$. Thus we deal with a finite horizon linear quadratic nonzero–sum deterministic difference game. We will work backwards in time according to the dynamic programming principle as to see how the optimal solutions depend on the current state. From stage k onwards the costs are

$$J_i(x(k),k) = \frac{1}{2}\sum_{j=k+1}^{K}[x'(j)Q_ix(j) + u_i'(j-1)R_{ii}u_i(j-1)].$$

Last step. Suppose x(K–1) is known to the players, then if the players play noncooperatively, the optimal costs can formally be written as

$$V_i^N(x(K-1),K-1) = \frac{1}{2}x'(K-1)M_i(K-1)x(K-1)$$

for appropriate matrices $M_i(K-1)$, i = 1, 2. If the players would play cooperatively, the optimal costs will formally be written as

$$V_i^C(x(K-1),K-1) = \frac{1}{2}x'(K-1)P_i(K-1)x(K-1)$$

for appropriate matrices $P_i(K-1)$, i = 1, 2. Cooperation is here for instance defined according to the cooperative solution I of section 3. If $V_i^C(x(K-1),K-1) \leq V_i^N(x(K-1),K-1)$, i = 1, 2, then the cooperative solution will be chosen and otherwise the noncooperative one. Whatever the decision, the costs of the last stage will formally be written as

$$V_i(x(K-1),K-1) = \frac{1}{2}x'(K-1)S_i(K-1)x(K-1),$$

where $S_i(K-1)$ equals either $M_i(K-1)$ or $P_i(K-1)$. Thus $S_i(K-1)$ becomes a function of x(K–1)! An exception is n = 1, i.e. x is a scalar, then $S_i(K-1)$ does not depend on x(K–1). For $n \geq 2$ $V_i(x(K-1),K-1)$ is piecewise quadratic in x(K–1), since $S_i(K-1)$ is piecewise constant in x(K–1). For n = 2, $S_i(K-1)$ is constant in cones of the x–space. For $n \geq 3$, the dependence of S(K–1) on x(K–1) becomes more complex.

One but last step. Suppose x(K–2) is known to the players. The cost functions to be considered now are

$$J_i(x(K-2),K-2) = \frac{1}{2}[x'(K-1)\{Q_i + S_i(K-1)\}x(K-1) + u_i'(K-2)R_{ii}u_i(K-2)],\ i = 1, 2,$$

in which we substitute $x(K-1) = Ax(K-2) + B_1u_1(K-2) + B_2u_2(K-2)$. Since $S_i(K-1)$ is state dependent for n > 1, the cost functions are no longer quadratic in the state; they are not even continuous in the state! In general, Nash solutions in pure strategies do not necessarily exist for discontinuous cost functions and one might then be forced to look for mixed solutions. This avenue has not been pursued any further.

References

[1] Basar, T., and G.J. Olsder, Dynamic noncooperative game theory, Academic Press, London, 1982.

[2] Fleming, W.H., and R.W. Rishel, Deterministic and stochastic optimal control, Springer Verlag, Berlin, 1975.

[3] Leitmann, G. (editor), Multicriteria decision making and differential games, Plenum Press, New York, 1976.

[4] Leitmann, G., Cooperative and noncooperative many player differential games, CISM Monograph nr. 190, Springer Verlag, Vienna, 1974.

[5] Olsder, G.J., Some thoughts about simple advertising models as differential games and the structure of coalitions, in: Directions in Large Scale Systems (Y.C. Ho and S.K. Mitter, eds.), Plenum Press, 1976.

[6] Olsder, G.J., Adaptive Nash strategies for repeated games resulting in Pareto solutions, Proceedings IFAC workshop on modelling, decisions and games with applications to social phenomena (J. Song and Y.C. Ho, eds.), 1986, pp. 447–454.

[7] Papavassilopoulos, G.P., J.V. Medanic and J.B. Cruz, On the existence of Nash strategies and solutions to coupled Riccati equations in linear–quadratic games, Journal of Optimization Theory and Applications, vol. 28, 1979, pp. 49–76.

[8] Papavassilopoulos, G.P., and G.J. Olsder, On the linear–quadratic, closed–loop, no–memory Nash game, Journal of Optimization Theory and Applications, vol. 42, 1984, pp. 551–560.

[9] Radner, R., Monitoring cooperative agreements in a repeated principal–agent relationship, Econometrica, vol. 49, 1981, pp. 1127–1148.

A COOPERATIVE DIFFERENTIAL GAME IN PREDATOR-PREY SYSTEMS

George Bojadziev
Department of Mathematics and Statistics
Simon Fraser University
Burnaby, B.C. Canada V5A 1S6

Abstract. A Lotka-Volterra predator-prey type model involving control is considered as a game between two players. While seeking conflicting growth objectives, the players design in cooperation a self-controlled growth policy aimed to establish a desirable population size level.

1. INTRODUCTION

The classical deterministic models in population dynamics usually predetermine the population behaviour. They do not reflect abrupt changes which occur in the real life due to diseases, climatic conditions, change of population size, etc. Also they do not take into consideration that in nature and society, besides the usual types of interaction (predator-prey interaction, competition, and cooperation), more complicated situations can exist.

The aim of this paper is mainly to focus attention on a relationship between populations characterized by predator-prey interaction combined with cooperation. The predator and the prey while seek conflicting growth objectives will design in cooperation a self-controlled growth policy in order both to reach and maintain a desirable population size level.

The study will be based on the differential game approach (see Blaquire, Gerard, and Leitmann [1], Isaacs [2], Szép and Forgó [3]) and on a Liapunov design avoidance control methodology developed by Leitmann and Skowronski [4] (see also Bojadziev and Skowronski [5] and Bojadziev [6]).

2. THE MODEL

We consider the Lotka-Volterra type predator-prey model with control

$$\dot{\underset{\sim}{x}}(t) = \underset{\sim}{f}(\underset{\sim}{x}(t), \underset{\sim}{u}(t)), \qquad \dot{\underset{\sim}{x}} = d\underset{\sim}{x}/dt, \qquad t \geq 0, \tag{1}$$

where $\underset{\sim}{x} = (x_1, x_2)^T \in R_+^2$, $\underset{\sim}{f} = (f_1, f_2)^T$, and

$$\begin{aligned} f_1(\underset{\sim}{x}, \underset{\sim}{u}) &= x_1(\alpha_1 - \beta_1 x_2) + u_1 x_1^2, \\ f_2(\underset{\sim}{x}, \underset{\sim}{u}) &= x_2(-\alpha_2 + \beta_2 x_1) + u_2 x_2^2. \end{aligned} \tag{2}$$

Here x_1 is the size of the prey, x_2 is the size of the predator, α_i and β_i, i=1,2, are positive constants, and $\underset{\sim}{u}(t) = (u_1,u_2)^T \in U \subset R^2$ is a control vector to be specified according to a growth restricted policy based on cooperation while x_2 preys on x_1. The biological meaning of the control terms $u_i x_i^2$, i=1,2, in (1) is that for $u_i > 0$ the growth of the i-th population is enhanced and for $u_i < 0$ it is dampered.

We denote the response of (1) by $\underset{\sim}{x}(t) = \underset{\sim}{k}(\underset{\sim}{x}(t_0), \underset{\sim}{u}, t)$ where $\underset{\sim}{x}(t_0) \in R_+^2$ is the initial state.

The nontrivial equilibrium of the controlled model (1), $\underset{\sim}{u}=0$, i.e. the Lotka-Volterra (L-V) model, is $E^o(x_1^o,x_2^o)$, where

$$x_1^o = \frac{\alpha_2}{\beta_2}, \qquad x_2^o = \frac{\alpha_1}{\beta_1}. \tag{3}$$

The equilibrium of the controlled model (1), assuming that

$$u_1u_2 + \beta_1\beta_2 \neq 0 \tag{4}$$

and dismissing the trivial equilibria on the axis x_1 and x_2, is $E^u(x_1^u,x_2^u)$, where

$$x_1^u = \frac{\beta_1\alpha_2-\alpha_1u_2}{u_1u_2+\beta_1\beta_2}, \qquad x_2^u = \frac{\beta_2\alpha_1+\alpha_2u_1}{u_1u_2+\beta_1\beta_2}. \tag{5}$$

The Jacobian determinant $J = x_1^u x_2^u(u_1u_2+\beta_1\beta_2)$ of the vector field of (1) calculated at E^u and fixed $\underset{\sim}{u} \in U$ is not zero according to (4), hence E^u is a simple equilibrium. The characteristic polynomial of (4) is $\lambda^2 - p\lambda + J$, where $p = u_1x_1 + u_2x_2$. If $p^2 - 4J < 0$, E^u is a focus; if $p^2 - 4J > 0$, E^u is a node $(J > 0)$ and saddle $(J < 0)$; if $p^2 - 4J = 0$, E^u is a node. The focus or node are asymptotically stable if $p < 0$, i.e. if

$$u_1x_1 + u_2x_2 < 0 \tag{6}$$

and unstable if $p > 0$.

3. THE GAME OBJECTIVES

We consider the predator and prey populations as two players who agree to exercise control on their own populations in order for both players to gain as a benefit a desirable population size level.

In the particular case of no control, $u_1 = u_2 = 0$, the classical L-V model (1) has response $\underset{\sim}{x}(t) = \underset{\sim}{k}(\underset{\sim}{x}(t_0),\underset{\sim}{0},t)$ which geometrically is represented by a closed orbit (periodic solution) in R_+^2 around the equilibrium (3) (Fig. 1). Depending on the initial state $\underset{\sim}{x}(t_0)$ the orbit may be close to the axis x_1 (small number of predator) or x_2 (small number of prey) which leads to large size fluctuations. Such a situation may be unacceptable to both populations (players).

In the general case, $\underset{\sim}{u} = \neq \underset{\sim}{0}$, each choice of control, say $\underset{\sim}{u}(t) = \underset{\sim}{c}_0$, $\underset{\sim}{c}_0 = (c_1^o, c_2^o)^T$, on some time interval starting at $t = t_0$, generates a solution or response $\underset{\sim}{k}(\underset{\sim}{x}(t_o), \underset{\sim}{c}_o, t)$ of the system (1) with initial state $\underset{\sim}{x}(t_o) \in R_+^2$ (geometrically represented by an orbit ℓ_0 in the phase plane (x_1, x_2)), and an equilibrium E^{c_0} obtained by (5). The response moving on the orbit ℓ_0 may or may not tend to E^{c_o}. A move of the response far away from E^{c_o} may endanger the existence of an acceptable size of either one or both populations. In order to avoid such situations, the populations may opt to self-control their own growth. This can be accomplished by changing the control value from $\underset{\sim}{c}_0$ to a suitable value $\underset{\sim}{c}_s$, $\underset{\sim}{c}_s = (c_1^s, c_2^s)^T$ at a switching point $\underset{\sim}{x}(t_s) = (x_1^s, x_2^s)^T$ on some time interval starting at $t = t_s$. As a result a response $\underset{\sim}{k}(\underset{\sim}{x}(t_s), c_s, t)$ along the orbit $\ell_s \neq \ell_o$ is generated which avoids entering of an undesirable region A (Fig. 1).

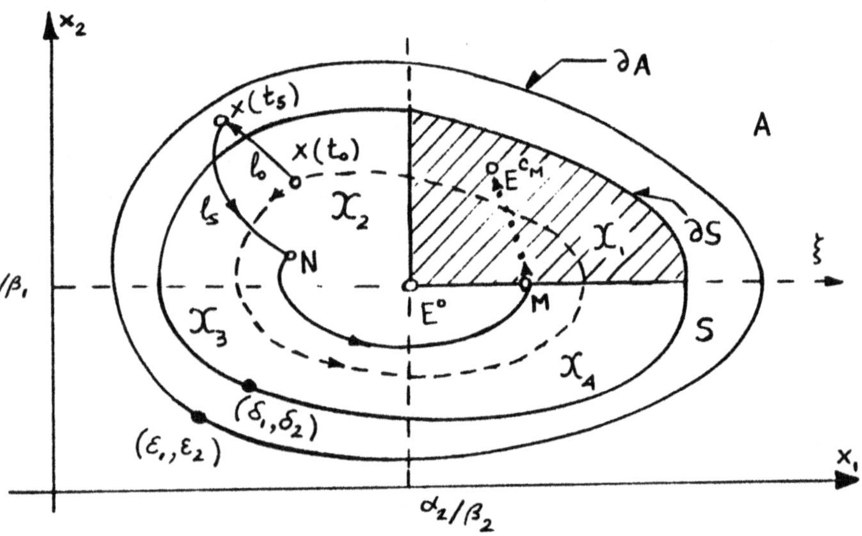

Fig. 1

However the achievement of avoidance may not be always a satisfactory solution for both or one of the populations. For example if the response is in region X_3, both populations have "small" size, in region X_2 the prey has a "small" size, the predator - "large" size, in region X_4

the prey has a "large" size, the predator - "small" size. The prey population has a large size in X_1 and X_4 while the predator population has a large size in X_1 and X_2. The players have the conflicting growth objectives to maintain their size in the above regions, correspondingly. However both players will have relatively large size level in region X_1 and may decide that this is a good reason for cooperation. On the basis of this observation we state the following objectives.

Control game objectives. The two players having conflicting growth objectives agree to cooperate by selecting a qualitative self-controlled program $\underset{\sim}{u}(t) \subset U$ the aim of which is to control the growth of each player in order both players to approach, reach, and maintain a size level in region X_1. In other words, the response of (1) should enter and stay in X_1 (reaching with capture).

In the next section we use a Liapunov function to introduce certain definitions and prove a theorem which will facilitate the design of the control program.

4. THE LIAPUNOV FUNCTION

The uncontrolled model (1) ($\underset{\sim}{u}=0$) has a Liapunov function

$$V(\underset{\sim}{x}) = \beta_2\left(x_1 - \frac{\alpha_2}{\beta_2}\right) - \alpha_2 \ln \frac{\beta_2 x_1}{\alpha_2} + \beta_1\left(x_2 - \frac{\alpha_1}{\beta_1}\right) - \alpha_1 \ln \frac{\beta_1 x_2}{\alpha_1} \tag{7}$$

with the following properties:

(i) The minimum of $V(\underset{\sim}{x})$ is attained at the equilibrium E^o given by (2);

(ii) $V(\underset{\sim}{x})$ is monotone increasing about E^o;

$V(x) \to \infty \ \|x\| \to 0, \qquad V(x) \to \infty$ as $\|x\| \to \infty$;

(iii) $$\frac{dV(\underset{\sim}{x})}{dt} = \sum_{i=1}^{2} \frac{\partial V}{\partial x_i} f_i(\underset{\sim}{x},\underset{\sim}{0}) = 0 , \tag{8}$$

where f_i are given by (2).

The function $V = V(\underset{\sim}{x})$ given by (7) describes in R^3 a "cup-shaped" surface with minimum E^o. The equation

$$V(\underset{\sim}{x}) = h , \qquad h = \text{const} > 0 \tag{9}$$

represents a family of level surfaces in R^3 whose orthogonal projection onto R^2 generates family of curves with the following properties: (a) they are orbits of the uncontrolled model (1); (b) do not intersect; (c) contain inside only the equilibrium E^o which is a centre; (d) if

$h_1 < h_2$, the curve generated by h_1 is inside of the curve generated by h_2.

5. AVOIDANCE CONTROL

Following papers [4,5,6] we introduce four definitions and prove a theorem

Definition 1. (Avoidance set A). Given $\underset{\sim}{\varepsilon} = (\varepsilon_1, \varepsilon_2)$, $\varepsilon_1, \varepsilon_2 > 0$ and $V(\underset{\sim}{x})$ by (7),

$$A \triangleq \{\underset{\sim}{x} \in R^2 : V(\underset{\sim}{x}) \geq V(\underset{\sim}{\varepsilon})\} . \tag{10}$$

The avoidance parameters $\varepsilon_1, \varepsilon_2$ can be small as desired for each particular case. At the boundary ∂A of A, $V(\underset{\sim}{x}) - V(\underset{\sim}{\varepsilon}) = 0$. The set A contains all anti-targets to be avoided (Fig. 1).

Definition 2 (Security zones). Given $\underset{\sim}{\delta} = (\delta_1, \delta_2)^T$, $\delta_i > \varepsilon_i$, and $V(\underset{\sim}{x})$ by (7),

$$S \triangleq \Delta_\delta \setminus A \neq \phi, \qquad \Delta_\delta \triangleq \{\underset{\sim}{x} \in R^2 : V(\underset{\sim}{x}) \geq V(\underset{\sim}{\delta})\} . \tag{11}$$

δ_i are security parameters. At the boundary ∂S of Δ_δ, $V(\underset{\sim}{x}) - V(\underset{\sim}{\delta}) = 0$. The set Δ_δ contains the security zone S and the avoidance region A. The allowable region for the motions of (1) is $R_+^2 \setminus A$.

Definition 3. (Desirable region Ω). $\Omega \triangleq R_+^2 \setminus \Delta_\delta$.

Definition 4. The set A defined by (10) is avoidable if there is a set $S \neq \phi$ given by (11) and a control $\underset{\sim}{u} \in U$ such that for all $\underset{\sim}{x}^s(t_s) \in S$, $\underset{\sim}{x}^s = (x_1^s, x_2^s)^T$, the response $\underset{\sim}{k}(\underset{\sim}{x}^s, \underset{\sim}{u}, t)$ of (1) cannot enter A, i.e.

$$\underset{\sim}{k}(\underset{\sim}{x}^s, \underset{\sim}{u}, t) \cap A = \phi \forall t . \tag{12}$$

Now using [4] we prove a theorem that gives sufficient conditions for the avoidance of A.

Theorem. The set A is avoidable if for $\underset{\sim}{x}(t)$ modelled by (1) there is a policy based on a control $\underset{\sim}{u}(t)$ and a C^1 function $\tilde{V}(\underset{\sim}{x})$ defined in an open set containing the security zone S such that

$$\text{(i)} \quad V(\underset{\sim}{x}^s) < V(\underset{\sim}{x}^a) , \qquad \underset{\sim}{x}^s \in S, \qquad \underset{\sim}{x}^a \in \partial A , \tag{13}$$

$$\text{(ii)} \quad \frac{d\tilde{V}(\underset{\sim}{x})}{dt} = \sum_{i=1}^{2} \frac{\partial \tilde{V}}{\partial x_i} f_i(\underset{\sim}{x}, \underset{\sim}{u}) \leq 0 . \tag{14}$$

Proof. Assume that A is not avoidable, i.e. (12) is violated. Hence for some $\underset{\sim}{x}^s(t_s) \in S$,

$$\underset{\sim}{k}(\underset{\sim}{x}^s, \underset{\sim}{u}, t) \cap A \neq \phi, \qquad t > t_s .$$

Then there is a $t_a > t_s$ for which $\underset{\sim}{x}^a(t_a) = \underset{\sim}{k}(\underset{\sim}{x}^s,\underset{\sim}{u},t_a) \in \partial A$. According to (13), $\tilde{V}(\underset{\sim}{x}^s(t_s)) < \tilde{V}(\underset{\sim}{x}^a(t_a))$, meaning that the function $\tilde{V}(\underset{\sim}{x})$ is increasing. This contradicts to (13) which states that $\tilde{V}(\underset{\sim}{x})$ is non-increasing along every solution to (1).

To design a policy for avoidance the region A by the response of (1) we use the Liapunov function $\tilde{V}(x)$ given by (7) and the above Theorem.

We observe that $V(\underset{\sim}{x})$ satisfies condition (13), i.e. $V(\underset{\sim}{x}^s) < V(\underset{\sim}{x}^a)$, $\underset{\sim}{x}^s \in S$, $\underset{\sim}{x}^a \in \partial A$. In order $V(x)$ to satisfy condition (14), we require the fulfillment of the condition

$$\sum_{i=1}^{2} \frac{\partial V(\underset{\sim}{x})}{\partial x_i} \cdot f_i(\underset{\sim}{x},\underset{\sim}{u}) \le 0 \ ,$$

which with (1) and (7) takes the form

$$\left(\beta_2 - \frac{\alpha_2}{x_1}\right) u_1 x_1^2 + \left(\beta_1 - \frac{\alpha_1}{x_2}\right) u_2 x_2^2 \le 0 \ . \tag{15}$$

This inequality establishes a relationship between the control $\underset{\sim}{u} = (u_1,u_2)^T$ and the population vector $\underset{\sim}{x}$.

6. REACHING WITH CAPTURE

According to the objectives stated in Section 3 the response of (1) must reach X_1 and stay there (reaching with capture). The region (target) X_1 is defined as

$$X_1 \triangleq \{\underset{\sim}{x}:\ \underset{\sim}{x}^u \in \Omega = R_+^2 \setminus \Delta_\delta,\ x_1^u > x_1^o,\ x_2^u > x_2^o\} \ . \tag{16}$$

For that purpose we require that the equilibrium $E^u(x_1^u,x_2^u)$ of (1) is located in X_1 and is asymptotically stable, hence it will attract the orbits of (1).

The conditions that $x_1^u > x_1^o$ and $x_2^u > x_2^o$ with (5) leads to

$$-\alpha_1\beta_2 u_2 > \alpha_2 u_1 \ , \qquad \beta_1\alpha_2 u_1 > \alpha_1 u_1 u_2 \ .$$

This inequalities restrain the control to the set

$$\left\{\underset{\sim}{u}:\ u_1 < -\frac{\alpha_1\beta_2}{\alpha_2}\ ,\ u_2 > \frac{\beta_1\alpha_2}{\alpha_1}\right\} \cup \{\underset{\sim}{u}:\ u_1 < 0,\ u_2 > 0\} \ . \tag{17}$$

The condition $\underset{\sim}{x}^u \in \Omega$ with the use of (7) and the security parameters (δ_1,δ_2) can be written as

$$V(\underset{\sim}{x}^u) < V(\underset{\sim}{\delta})$$

or after some calculations as

$$\beta_2(x_1^u-\delta_1) - \alpha_2 \ln \frac{x_1^u}{\delta_1} \beta_1(x_2^u-\delta_2) - \alpha_1 \ln \frac{x_2^u}{\delta_2} < 0 . \tag{18}$$

The condition for asymptotic stability of E^u is found in Section 2 and is given by (6).

Now we finally specified the set U as the intersection of the sets: (a) (15) - condition for avoidance of A; (b) (17) and (18) - conditions that $E^u \in X_1$; (c) (6) - condition that E^u is asymptotically stable equilibrium.

7. GAME CONTROL PROGRAM

We can assume that the response of (1), $\underset{\sim}{k}[t] = \underset{\sim}{k}[\underset{\sim}{x}(t_1), \underset{\sim}{u}(t_1), t]$, $t > t_1$ $\underset{\sim}{x}(t_1)$ is the initial state, $u(t_1) \in U$ is a fixed control, belongs to Ω. Otherwise if $\underset{\sim}{k}[t]$ enters the security zone S given by (11), a new control $\underset{\sim}{u}(t_s)$, $t_s > t_1$, can be selected from U at a switching point $\underset{\sim}{x}(t_s) \in S$ with corresponding response $\underset{\sim}{k}[\underset{\sim}{x}(t_s), \underset{\sim}{u}(t_s), t]$ along an orbit going into Ω.

If the response is in the region X_1, any control selected from U will keep the response in X_1. If the response is in the regions X_2, X_3, or X_4, null control $(u=0)$ have to be used (on Fig. 1 null control is applied at the point $N \in X_2$). Then the response of (1) moves along an orbit of the L-V model with centre E^o until reaches a point M with coordinates $x_1^M > x_1^o$, $x_2^M = x_2^o$ on the semi-axis $E^o\xi_1$ (Fig. 1). At that point a control $u = c_M \in U$ is applied so that the response approaches asymptotically the stable equilibrium $E^{c_M}(x_1^{c_M}, x_2^{c_M}) \in X_1$, hence $\underset{\sim}{k}[t]$ is captured in X_1.

According to the mutually negotiated cooperation policy and pursuing their conflicting growth objectives, the players x_i, $i=1,2$, have the option to select the control $\underset{\sim}{u} = c_M$ in such a fashion that one of the populations will benefit more or less than the another population (player).

REFERENCES

1. Blaquire, A., F. Gerard, and G. Leitmann. Quantitative and Qualitative Games. Academic Press, New York, 1969.

2. Isaacs, R. Differential Games. Wiley, New York, 1964.

3. Szép, J. and F. Forgó. Introduction to the theory of games. D. Reidel Publishing Company, Boston, 1985.

4. Leitmann, G. and J. Skowronski. Avoidance control. J. Optim. Theory and Appl., 23, 581-591 (1977).

5. Bojadziev, G. and J. Skowronski. Controlled food consumption. Methods of Operations Research, 49, 499-506 (1985).

6. Bojadziev, G. Behavioural strategy of some controlled predator-prey systems. Advances in Optimization and Control, Proceeding, Montreal, Canada, 1986. Lecture Notes 301, Springer-Verlag, 283-292.

REPEATED BARGAINING UNDER UNCERTAINTY

Harri Ehtamo, Jukka Ruusunen and Raimo P. Hämäläinen
Espoo, Finland

Abstract : The Nash bargaining scheme is extended to discrete and continuous time dynamic games. The bargaining takes place under future uncertainty and the contracting takes place at each stage on the basis of the available information. A two-level hierarchical method to handle such problems is presented. At the lower level a weighted sum of the decision makers' objectives is maximized. This corresponds to solving the cooperative problem with a given contract, which is expressed by the weights of the individual objectives. At the top level the contract is determined by solving an algebraic equation such that the rationality axioms of the Nash bargaining scheme are satisfied for the whole time interval of cooperation.

1. Introduction

In this paper we shall consider a situation, where a group of decision makers (DMs) acts cooperatively. It is assumed that the interests of the individual DMs are conflicting. If the DMs pursued their own interests, a noncooperative status quo outcome would result. The status quo outcome can be e.g. a noncooperative Nash outcome. On the other hand, if the DMs act cooperatively they select policies that consider the interests of all the DMs. One way of selecting a fair cooperative outcome is the axiomatic approach to bargaining. In the Nash bargaining model a solution to the cooperative game is the unique outcome that maximizes the geometric average of the DMs' gains from cooperation. The gains are measured relative to the status quo outcome.

A continuous time, "open-loop" bargaining game with fixed initial state is studied in Section 2. Since the Nash bargaining solution has been developed for games in normal form, see e.g. Roth (1979), the generalization to the open-loop games is straightforward. Necessary conditions to the bargaining solution are given and it is shown that a convenient method to calculate the bargaining solution is to transform the original problem to a two-level hierarchical decision problem. At the lower level of the hierarchy a weighted sum of the DMs' objectives is maximized. At the top level the contract is determined by solving an algebraic equation such that the rationality axioms of the Nash bargaining scheme are satisfied. For the applications of this method see Ehtamo *et al* (1986, 1989).

In the next section it is shown that the hierarchical method can be applied to more general cases, where the decision spaces are arbitrary sets. In fact, it is shown that the Nash bargaining problem is equivalent to a two-level hierarchical decision problem.

In Section 4 bargaining takes place in a dynamic framework under future uncertainty, and the contracting takes place at each stage on the basis of the available information. The gains are measured by considering the past, the current and the expected future gains. The hierarchical method gives a computational procedure for repeated bargaining under uncertainty. Conequently the Pareto optimal bargaining policy parametrized by the weight vector μ can be calculated beforehand, e.g. by using dynamic programming. The contract μ is then chosen at each stage on the basis of new information and so that the rules of fairness of the Nash bargaining scheme hold for the whole time interval of cooperation.

2. An example of the Nash bargaining problem

In this section we shall consider the dynamic system

$$\dot{x}(t) = f(x(t), u(t), t), \quad x(t_0) = x_0, \quad t \in [t_0, t_f], \tag{1}$$

and the payoffs

$$J_i(u) = h_i(x(t_f)) + \int_{t_0}^{t_f} g_i(x(t), u(t), t)dt, \quad 1 \leq i \leq M, \tag{2}$$

where i refers to DMi; $x(t) \in R^n$, $u \hat{=} (u_1, \ldots, u_M)$, where $u_i(t) \in R^{m_i}$, and the functions f, h_i, g_i are in C^1 with respect to x, u and continuous in t. An admissible decision u_i for DMi is assumed to be a piecewise continuous function satisfying a constraint of the form

$$k_i(u_i(t), t) \geq 0, \tag{3}$$

where $k_i : R^{m_i} \times R \rightarrow R^{q_i}$ is differentiable in u_i. The set of all such decisions is denoted by Ω_i, $1 \leq i \leq M$, and the admissible set $\Omega_1 \times \ldots \times \Omega_M$ is denoted by Ω. The initial state $x(t_0) = x_0$ and the time interval $[t_0, t_f]$ are fixed.

The DMs are assumed to follow the Nash's bargaining scheme (see e.g. Roth (1979)). Hence we consider the following problem, called BP:

$$\max \prod_{i=1}^{M} \Big(J_i(u) - D_i\Big) \tag{4}$$

$$\text{subject to } u \in \Omega, \ J_i(u) \geq D_i \text{ for all } i, \tag{5}$$

where D_i is a status quo outcome for DMi. $(D_1, \ldots, D_M)$ can be defined e.g. as a noncooperative Nash outcome, provided such exists. Define

$$\dot{z}_i(t) = g_i(x(t), u(t), t), \quad z_i(t_0) = 0, \quad t \in [t_0, t_f], \quad 1 \leq i \leq M. \tag{6}$$

Then BP is equivalent to

$$\begin{aligned} &\max \prod_{i=1}^{M} \Big(z_i(t_f) + h_i(x(t_f)) - D_i\Big) \\ &\text{subject to (1), (5), (6).} \end{aligned} \tag{7}$$

Set

$$H(x,u,\lambda,\mu,t) \triangleq \sum_{i=1}^{M} \mu_i\, g_i(x,u,t) \;+\; \lambda'\, f(x,u,t). \tag{8}$$

Suppose there is at least one point $u \in \Omega$ such that $J_i(u) > D_i$ for all i and suppose u^* solves (7). Then u^* satisfies the following equations:

$$\dot{x}^* = f(x^*,u^*,t), \quad x^*(t_0) = x_0, \tag{9}$$

$$\dot{z}_i^* = g_i(x^*,u^*,t), \quad z_i^*(t_0) = 0, \quad 1 \le i \le M, \tag{10}$$

$$\dot{\lambda}' = -H_x(x^*,u^*,\lambda,\mu,t),$$

$$\lambda'(t_f) = \sum_{i=1}^{M} \mu_i\, h_{ix}(x^*(t_f)), \tag{11}$$

$$\dot{\mu}_i = 0,$$

$$\mu_i(t_f) = \prod_{\substack{j=1\\ j\neq i}}^{M} \Big(z_j^*(t_f) + h_j(x^*(t_f)) - d_j \Big), \quad 1 \le i \le M, \tag{12}$$

$$H(x^*(t),u^*(t),\lambda(t),\mu(t),t) \ge H(x^*(t),u(t),\lambda(t),\mu(t),t) \quad \text{for all } u \in \Omega, \text{ and } t \in [t_0,t_f]. \tag{13}$$

Equations (9)–(13) are the maximum principle necessary conditions for u^* to solve BP. Hence, the solution candidates to BP can be found by solving the resulting $2(n+M)$–dimensional two-point boundary value problem. Although efficient numerical methods have been developed to solve such problems, it is convenient to do this, especially if M is large, by using the following method. First note that μ in (11)–(13) is a strictly positive constant. For

$$\mu \in \Lambda^{>} = \{\lambda \in R^M \mid \lambda_i > 0, \ \forall i\} \tag{14}$$

let $\omega(\mu)$ be a solution to the $2n$–dimensional two-point boundary value problem associated with (9), (11) and (13). Set

$$T_i(\mu) = \prod_{j\neq i} \Big(J_j(\omega(\mu)) - D_j \Big), \ 1 \le i \le M, \tag{15}$$

$$T(\mu) = [T_1(\mu), \ldots, T_M(\mu)]'. \tag{16}$$

Then $T : \Lambda^{>} \to R^M$. If $\mu^* \in \Lambda^{>}$ solves the equation

$$T(\mu) - \mu = 0, \tag{17}$$

then the corresponding $u^* = \omega(\mu^*)$ solves (9)–(13). Equation (17) is solved by using an appropriate iterative method. Numerical experiments have shown, see Ehtamo *et al* (1987, 1989), that an effective algorithm here is e.g. the quasi-Newton method of Broyden. The algorithm to find u^* proceeds as follows:

Step 1: Set $k = 0$ and choose a starting vector $\mu^0 \in \Lambda^{>}$.

Step 2: For $\mu^k \in \Lambda^>$ compute $\omega(\mu^k)$ and $T(\mu^k)$.

Step 3: If $\| T(\mu^k) - \mu^k \| < \epsilon$, set $\mu^* = \mu^k$. Else, choose the next trial point $\mu^{k+1} \in \Lambda^>$, set $k := k+1$, and go to Step 2.

3. The general case

In this section we shall show that the algorithm presented in the previous section is also applicable to more general problems.

Let Ω_i be the decision set for DMi, $1 \le i \le M$, set $\Omega = \Omega_1 \times \ldots \times \Omega_M$, and let J_i, $J_i : \Omega \to R$, be the objective functional for DMi. Set

$$F_i(u) = J_i(u) - D_i, \quad 1 \le i \le M, \tag{18}$$

$$F(u) = \prod_{i=1}^{M} F_i(u), \quad u \in \Omega, \tag{19}$$

where $D_i \in R$ is a status quo outcome for DMi. Consider problem BP$'$

$$\max_{u \in \Omega} F(u) \tag{20}$$

$$\text{subject to } F_i(u) \ge 0, \quad 1 \le i \le M. \tag{21}$$

The following holds [for the proofs of the lemmas see Ehtamo *et al* (1987)]:

Lemma 1: Let $u^* \in \Omega$ be such that

$$\sum_{i=1}^{M} \mu_i F_i(u) \le \sum_{i=1}^{M} \mu_i F_i(u^*) \quad \text{for all } u \in \Omega, \tag{22}$$

where

$$\mu_i = \prod_{j \ne i} F_j(u^*), \; \mu_i > 0, \; 1 \le i \le M. \tag{23}$$

Then u^* solves BP$'$.

Remark 1: Note that $\mu_i > 0$ for all i implies $F_i(u^*) > 0$ for all i. Namely, from (22), (23) it follows that $F_k(u^*) > 0$ for some k (provided there is $u \in \Omega$ such that $F_i(u) \ge 0$ for all i). Since $F_i(u^*)\mu_i = F(u^*) = F_k(u^*)\mu_k > 0$ for all i, it follows that $F_i(u^*) > 0$ for all i.

Define

$$S = \{(F_1(u), \ldots, F_M(u)) \mid u \in \Omega\} \subset R^M, \tag{24}$$

let $\Lambda^{\le} \subset R^M$ be the convex cone $\Lambda^{\le} = \{\lambda \in R^M \mid \lambda_i \le 0, \forall i\}$, and set $Y = S + \Lambda^{\le}$. Suppose that BP$'$ has a solution $u^* \in \Omega$ (observe that BP$'$ always has a solution $u^* \in \Omega$ if $\{y \in Y \mid y_i \ge 0, \forall i\}$ is compact, since the geometric average is a continuous function). Then the corresponding optimal outcome $(J_1(u^*), \dots, J_M(u^*))$ is *Pareto optimal*. Moreover, if Y is a convex set, then the optimal outcome $(J_1^*, \dots, J_M^*)$ is *unique*, and we have

Lemma 2: Let Y be a convex set, and let $u^* \in \Omega$ solve BP′ with $F_i(u^*) > 0$ for all i. Then (22) holds, where μ_i is as in (23).

Remark 2: It can be shown that if Ω is a convex set in a vector space U, and $F_i : \Omega \to R$, $1 \le i \le M$, are concave functionals on Ω, then Y is convex.

According to Lemma 1 BP′ can be solved by using the algorithm of Sect. 2, provided $\omega(\mu^k)$ is defined as a solution to the problem

$$\max_{u \in \Omega} \sum_{i=1}^{M} \mu_i^k J_i(u). \tag{25}$$

4. An adaptive contract

In this section we consider a dynamic bargaining model in a stochastic environment. The dynamic character of this model is due to the stage by stage bargaining. However, the results given can be generalized to more complex problems of stochastic control, e.g. to problems involving dynamic systems.

Consider a group $\theta_1 = \{1, \dots, M\}$ of M DMs who are in a position to cooperate. The cooperation takes place over N time periods. In each period the outcome of the game is affected by a random disturbance ξ_k, $k \in \theta_2 = \{0, \dots, N-1\}$. ξ_k is an element of a set Ξ_k, and it is characterized by a probability measure $p_k(\cdot|I_{k-1})$, where $I_{k-1} = (\xi_{-1}, \xi_0, \dots, \xi_{k-1})$ is the information related to period $k-1$. ξ_{-1} describes the past behaviour of the random variable before the initial period $k = 0$. It is assumed that the actual value taken by ξ_k is known before the decision in period k is made.

The admissible policies are assumed to be functions of the information vector and such that $u_k(I_k) \in U_k \subset R^m$ for all I_k, $k \in \theta_2$. Furthermore, u_k is of the form $u_k = (u_k^1, \dots, u_k^M)$, where $u_k^i \in U_k^i \subset R^{m_i}$ is the admissible policy for DMi, $i \in \theta_1$. The decision u_k and the sequence of random variables described by the information I_k yield a payoff $g_k^i(u_k, I_k)$ for DMi in period k. The objective of DMi is the maximization of the sum of these payoffs over the time interval of cooperation,

$$J^i = \sum_{l=0}^{N-1} g_l^i(u_l, I_l). \tag{26}$$

The benefits of cooperation are measured from a status quo, which would prevail in the noncooperative situation. The status quo policies are denoted by $v_k = \delta_k(I_k)$, $k \in \theta_2$.

The problem is to define a contract related to period k such that the product of the individual gains is maximized over the *whole* time interval of cooperation. Consequently, the time interval is partitioned into the past and the future relative to each period $k \in \theta_2$. The past periods are described by the cumulative benefits for the DMs up to period k

$$\begin{aligned} y_k^i &= y_{k-1}^i + g_{k-1}^i(u_{k-1}, I_{k-1}) - g_{k-1}^i(\delta_{k-1}(I_{k-1}), I_{k-1}), \\ y_0^i &= 0. \end{aligned} \tag{27}$$

The expected future payoffs from period k on are

$$J_k^i(u_k, \dots, u_{N-1}; I_k) = g_k^i(u_k(I_k), I_k) + E_{\xi_{k+1}, \dots, \xi_{N-1}} \{ \sum_{l=k+1}^{N-1} g_l^i(u_l(I_l), I_l) \}. \tag{28}$$

For the status quo policies these expected future payoffs are denoted by

$$D_k^i(I_k) = J_k^i(v_k, \dots, v_{N-1}; I_k). \tag{29}$$

The contract definition problem related to period $k \in \theta_2$ is now stated as follows: For given y_k^i, I_k find an admissible policy $\{\bar{u}_k, \dots, \bar{u}_{N-1}\}$ such that

$$\begin{aligned} &\prod_{i \in \theta_1} \left[y_k^i + J_k^i(\bar{u}_k, \dots, \bar{u}_{N-1}; I_k) - D_k^i(I_k) \right] \\ &\qquad \geq \prod_{i \in \theta_1} \left[y_k^i + J_k^i(u_k, \dots, u_{N-1}; I_k) - D_k^i(I_k) \right] \end{aligned} \tag{30}$$

for all admissible policies $\{u_k, \dots, u_{N-1}\}$ for which

$$y_k^i + J_k^i(u_k, \dots, u_{N-1}; I_k) - D_k^i(I_k) \geq 0, \quad \forall i \in \theta_1. \tag{31}$$

For $l \in \theta_2$, $\mu = (\mu^1, \dots, \mu^M) \in \Lambda^>$ let $\omega_l(\mu)(I_l)$ solve the maximization problem

$$\max_{u_l \in U_l} \sum_{i \in \theta_1} \mu^i g_l^i(u_l, I_l). \tag{32}$$

Then, for $\mu \in \Lambda^>$,

$$\begin{aligned} &\sum_{i \in \theta_1} \mu^i J_k^i(\omega_k(\mu), \dots, \omega_{N-1}(\mu); I_k) \\ &\qquad \geq \sum_{i \in \theta_1} \mu^i J_k^i(u_k, \dots, u_{N-1}; I_k) \end{aligned} \tag{33}$$

for all admissible $\{u_k, \dots, u_{N-1}\}$. For given $y_k = (y_k^1, \dots, y_k^M)$, I_k, let $\bar{\mu} \in \Lambda^>$ satisfy

$$\bar{\mu}^i = \prod_{\substack{j \in \theta_1 \\ j \neq i}} [y_k^j + J_k^j(\omega_k(\bar{\mu}), \dots, \omega_{N-1}(\bar{\mu}); I_k) - D_k^j(I_k)], \quad i \in \theta_1, \tag{34}$$

and set $\bar{u}_l = \omega_l(\bar{\mu})$ for $l \geq k$. Then it follows from Lemma 1 that $\{\bar{u}_k, \dots, \bar{u}_{N-1}\}$ solves (30) and (31). We then define

Definition 1: The admissible policy $\{\omega_0(\mu),\ldots,\omega_{N-1}(\mu)\}$ is the solution to the N–stage bargaining problem defined above if at stage k, $k \in \theta_2$, the contract μ is defined *adaptively* according to the Nash bargaining scheme.

Definition 1 gives rise to some comments. Since (32) and (34) define policies of the form $\bar{\gamma}_k(y_k, I_k)$, $k \in \theta_2$, see Ruusunen *et al* (1988), one may define admissible "feedback" bargaining policies by considering all the functions of the form $\gamma_k(y_k, I_k)$, $k \in \theta_2$, and consider maximization of

$$\prod_{i \in \theta_1} [y_k^i + J_k^i(\gamma_k,\ldots,\gamma_{N-1}; I_k) - D_k^i(I_k)], \tag{35}$$

subject to

$$y_k^i + J_k^i(\gamma_k,\ldots,\gamma_{N-1}; I_k) - D_k^i(I_k) \geq 0, \quad \forall i \in \theta_1, \tag{36}$$

where $J_k^i(\cdot;\cdot)$ is defined as in (28) but with u_l replaced by γ_l and the expectation calculated by taking into account the state equation (27). If $\{\gamma_0,\ldots,\gamma_{N-1}\}$ solves this problem it is called a *feedback bargaining solution* (FBS). The existence of the FBS is not clear, since the product of the individual DMs' objectives is not additive in k. For example, if one tries to calculate the FBS stage by stage backwards in time assuming that bargaining also takes place in the future, then the feasible set at stage k will be smaller than the original feasible set. Consequently, the policy obtained need not be maximizing, not even Pareto optimal. However, as one might expect, the following result can be shown to hold: If $J_k^i(u_k,\ldots,u_{N-1}; I_k)$ is a concave functional in an appropriate convex set Ω_k^i for all I_k, $i \in \theta_1$, $k \in \theta_2$, and if the FBS exists, then it is given by $\{\bar{\gamma}_0,\ldots,\bar{\gamma}_{N-1}\}$.

Finally, it should be noted that Definition 1, in addition to extending the Nash bargaining scheme, gives a computational procedure to multi-stage bargaining under uncertainty. The policy $\{\omega_0(\mu),\ldots,\omega_{N-1}(\mu)\}$ can be calculated beforehand, for example, by using the dynamic programming. The contract μ is then chosen at each stage on the basis of new information and so that the rules of fairness of the Nash bargaining scheme hold for the whole time interval of cooperation.

5. Conclusion

In this paper we have presented a survey of possible extensions of the Nash bargaining scheme to dynamic decision problems. In Section 2 a cooperative differential game was studied. The game was solved at a fixed initial state. The bargaining problem was transformed into a two–level hierarchical decision problem. At the lower level a weighted sum of the DMs' objectives is maximized. At the top level the contract is determined by solving an algebraic equation such that the rationality axioms of the Nash bargaining scheme are satisfied. The solution can be supposed to exist if there are points strictly dominating the status quo outcome. In Section 3 it was shown that the hierarchical algorithm can also be applied in a more general context.

In Section 4 multi–stage bargaining under uncertainty was studied. An adaptive mechanism that utilizes the observations about the random disturbances was presented. At each stage the product of the individual gains was maximized over the whole time interval of cooperation. The most efficient way of obtaining the bargaining solution is the proposed hierarchical algorithm. Again, the solutions can be found, provided there are outcomes at each stage dominating the status quo outcome. This is a very mild condition, since the dominance is required to hold over the whole time interval of cooperation, which also includes the past periods.

6. References

Ehtamo, H., J. Ruusunen, V. Kaitala and R. P. Hämäläinen (1986), "Solution for a dynamic bargaining problem with an application to resource management," Journal of Optimization Theory and Applications, Vol. 59, to appear.

Ehtamo, H., J. Ruusunen, and R. P. Hämäläinen (1987), "On the computation of the Nash bargaining solution with an energy management example," in Proc. 26th IEEE CDC, Los Angeles, CA, pp. 263–266.

Ruusunen, J., H. Ehtamo, and R. P. Hämäläinen (1988), "Bargaining with dynamic information," in Proc. 3rd Bellman Continuum, Springer–Verlag, to appear.

Ehtamo, H., J. Ruusunen, and R. P. Hämäläinen (1989), "A hierarchical approach to bargaining in power pool management," IEEE Transactions on Automatic Control, to appear.

Roth, A. E. (1979), "Axiomatic Models of Bargaining," Springer–Verlag, Berlin Heidelberg.

THE THEORY OF TEAMS: A SELECTIVE ANNOTATED BIBLIOGRAPHY*

TAMER BAŞAR and RAJESH BANSAL
Decision and Control Laboratory
Coordinated Science Laboratory
University of Illinois
1101 W. Springfield Avenue
Urbana, Illinois 61801/USA

ABSTRACT

This is an annotated bibliography of the literature on the theory of teams, starting with the pioneering work of Marschak and Radner in the early 1950's. In drawing papers from the vast literature in this area we have been rather selective, concentrating on stochastic models rather than deterministic, and highlighting those papers which have made (or, for more recent publications, have the potential to make) major impact on research in this area.

KEY WORDS & PHRASES. *Teams, decentralized stochastic control, decision making under uncertainty, person-by-person optimality, information patterns, algorithms and their complexity.*

A *team* is defined as a group of agents who work together in a coordinated effort, in an uncertain and possibly hostile environment, in order to achieve a common goal. In achieving this goal, members of the team, so-called *agents* or *decision makers*, do not necessarily acquire the same data or information, and hence they have to operate in a *decentralized* mode of decision making while optimizing a common objective functional. *Team theory* is the study of such situations, with applications in economics, control and communications.

The theory of teams originated in mathematical economics in the 1950's and early 1960's, when Marschak and Radner developed the concepts for the theory of a firm. The implications for, and relations with control theory were just beginning to emerge when Witsenhausen (1968) recognized the complications caused by nonclassical information structures for very simple stochastic teams. Team theory formally entered the control engineering literature through the works of Ho, Chu and others in the early 1970's. As its relationship with decentralized control theory and large scale system theory became increasingly clear, team decision theory (alongwith the theory of information structures) developed through the 1970's and 1980's. This compilation serves to introduce the reader to existing results in team theory, from mostly a control-theoretic point of view.

In order to keep the number of papers covered here at a reasonable level, we had to be selective in this bibliography. Our concentration has been on stochastic models, and theoretical developments rather than applications of team theory. From the past, we have included those papers which we feel have made major impacts on future work in the area; for more recent publications, our selection was guided by potential impact on future research. In the process of this compilation, we have also left out two large groups of papers, of considerable relevance to research on team theory. One of these groups is stochastic control, which can be viewed as single

*Compilation of this bibliography was completed during the period when our research on stochastic teams was supported in part by a Grant AFOSR 84 0056 from the US Air Force Office of Scientific Research.

agent team theory; here we have included only a few major references, as they relate to teams, and also the work on decentralized information patterns. The second group is nonzero-sum games, which is in a sense extension of the team framework to include multiple criteria scenarios (a different objective function for each agent). There is again a vast literature in this area, and we would be doing injustice to it by including just a few such references here; accordingly, we have left out even the key papers on nonzero-sum games unless there is a strong connection with team theory, such as the cases of Stackelberg games and some incentive design problems. Regarding the latter class of problems, that is, incentives and organizational design, though it has some common elements with team theory, it generally involves multiple criteria, and is approached from a different perspective. Hence this bibliography should not be considered as one also covering the field of organization design and incentives; we have included only those references where team-theoretic ideas and techniques play a major role.

The list to follow is organized in chronological order, with each paper given a three digit number just preceding the name of the first author; for example, [70-2] indicates that that particular paper is the second one included here among those published in 1970.

[54-1] **Marschak, J.** and **R. Radner**, "The firm as a team." (Abstract) *Econometrica*, 22, 523, 1954.

It was recognized that the firm could be regarded as a "team," i.e., a group of decision makers with a common goal, with decisions based on information about random "nature" variables, and the information being different for different agents. It was noted that the cost of information needs to be taken into account when performing utility maximization.

[55-1] **Marschak, J.**, "Elements for a theory of teams," *Management Science*, 1, 127-137, 1955.

A team was defined as a group of individuals each of whom takes decisions about something different but who receives a common reward based on all the decisions. It was noted that while better information leads to a higher group success score, a larger expense is incurred; i.e., one needs to allow for the cost of information. It was also noted that a realistic theory of teams needs to be dynamic.

[58-1] **Beckman, M. J.**, "Decision and Team Problems in Airline Reservations," *Econometrica*, 26, 134-145, 1958.

A reservation system charged with selling a fixed quantity of a highly perishable commodity through many agents is used as an instructive example of decision making and communication in teams/organizational theory. The example is based on Γ-distributions which adequately fit the probability distributions of late cancellations, no-shows and the number of standby passengers.

[59-1] **Radner, R.**, "Application of linear programming to team decision problems," *Management Science*, 5, 143-150. 1959.

Along the lines of his other paper, "The linear team: An example of linear programming under uncertainty," in *Proc. of the Second Symposium in Linear Programming*, Washington, DC, National Bureau of Standards (1955), Radner shows by means of an example how linear programming can be applied to obtain optimum team decision functions for the case where the payoff is a convex polyhedral function of the decision variables.

[59-2] **Marschak, T.**, "Centralization and decentralization in economic organizations," *Econometrica*, 27, 399-430, 1959.

The problem of choosing among several alternative schemes exhibiting various degrees of centralization is addressed. A class of schemes for organizational decision making including centralized, decentralized and unrestricted subclasses is considered. Criteria for ranking the schemes are obtained and it is found that a general preference for one of the subclasses cannot be defended unless the model is restricted substantially.

[61-1] **McGuire, C. B.**, "Some team models of a sales organization," *Management Science*, 7, 101-130, 1961.

A simplified mathematical model which deals with the day to day problems faced by a bakery sales force in selling a single product is considered. The solution is obtained in terms of a linear program.

[61-2] **Radner, R.**, "The evaluation of information in organizations," *Proc. of the 4th Berkeley Symposium on Probability and Mathematical Statistics*, University of California Press, Berkeley, CA, 491-530, 1961.

Certain elementary information structures for teams are described, evaluated and compared, so as to obtain a basis for comparison of general information structures. The discussion is restricted to the case where the payoff is a quadratic function of the action variables.

[62-1] **Radner, R.**, "Team decision problems," *Ann. Math. Stat.*, 33, 857-881, 1962.

This often referenced work gives a general formulation of the team decision problem and analyzes it from both the minimax and Bayes points of view. Explicit solutions are obtained for the quadratic payoff case. A most important result, regarding the linearity of the optimal decision functions in case the *a priori* distributions induce a normal distribution of the information variables, is also established.

[68-1] **Witsenhausen, H. S.**, "A counterexample in stochastic optimal control," *SIAM J. of Control*, 6, 131-147, 1968.

It is shown that under decentralized information nonlinear controllers may outperform the optimum linear controllers even when the objective is to minimize the expectation of a quadratic criterion for a linear system subject to Gaussian noise with unconstrained control variables. The paper marks the beginning of a large volume of work on information structures, and on how the nonclassical nature of the information pattern (arising due to memory limitations, delay or operating costs, etc.) contributes towards making a stochastic control or team problem intractable.

[68-2] **Wonham, W. M.**, "On the separation theorem of stochastic control," *SIAM J. Control*, 6, 312-326, 1968.

Using dynamic programming, it is shown that the combined control and filtering problem for a continuous stochastic linear dynamic system with classical information can be reduced to two independent problems of control and filtering, respectively, the result holding even if the cost is nonquadratic and the optimal controllers are nonlinear.

[68-3] **Marschak, J. and K. Miyasawa**, "Economic comparability of information systems," *International Economic Review*, 9, 137-174, 1968.

A partial order is defined on the set of information systems. One system is more informative than another if the expected payoff is larger under this system than under the other. The comparisons involve garbling (which includes collapsing of information) and joining information, and measure theoretic difficulties are avoided by confining to a finite set of events and messages.

[70-1] **Ho, Y. C.**, "Differential games, dynamic optimization and generalized control theory," Survey Paper, *JOTA*, 6, 179-209, 1970.

The problem of control is set in a generalized framework in which there are more than one criterion and more than one intelligent controller each of which has access to different information. It is argued that optimal control, differential games, dynamic team theory, etc. can all be viewed as special cases of this generalized control theory.

[71-1] **Witsenhausen, H. S.**, "Separation of estimation and control for discrete time systems," *Proc. of IEEE*, 59, 1557-1566, 1971.

A systematic formulation of decentralized stochastic teams is provided within a general framework, including a study of various information patterns that may arise in this context. The separation properties between estimation and control under various assumptions about the underlying system and the information pattern are examined, and the optimality of affine solutions for the LQG dynamic team problem under the one step delay pattern is asserted.

[71-2] **Chong, C. Y. and M. Athans**, "On the stochastic control of linear systems with different information sets," *IEEE Trans. on Automat. Contr.*, AC-16, 423-430, 1971.

It is shown that the separation theorem does not hold for general linear systems when the information available to distinct controllers is different. Thus, limitations on information exchange imply an increase in offline computational effort.

[71-3] **Witsenhausen, H. S.**, "On information structures, feedback and causality," *SIAM J. on Control*, 9, 149-160, 1971.

A contribution towards a systematic information structure theory, the paper brings the relation between causality and the existence of a unique solution into focus, by characterizing the game using its interactions with the policies of the agents rather than chronological order.

[71-4] **Chu, K. C. and Y. C. Ho**, "On the generalized linear-quadratic-Gaussian problems," in *Differential Games and Related Topics*, H. W. Kuhn and G. P. Szegö, Eds., Amsterdam: North Holland, 373-388, 1971.

An informal paper which addresses and provides partial answers to the question "When does the generalized linear-quadratic-Gaussian control problem have an optimal solution, and when is this solution linear?"

[72-1] **Ho, Y. C. and K. C. Chu**, "Team decision theory and information structures in optimal control problems - Parts I and II," *IEEE Trans. on Automat. Contr.*, AC-17, 15-22, 22-28, 1972.

Information structures of organizations are applied to dynamic team decision problems. The information structure is partially nested if decision makers in the same precedence tree can infer the information available to their precedents. Equivalence relations in information and in control functions among different team problems with partially nested information structure are developed, and these are shown to aid in solving many general problems by relating their solution to those of systems with perfect memory.

[72-2] **Marschak, J.** and **R. Radner**, *Economic Theory of Teams*, New Haven, CT: Yale University Press, 1972.

One of the few books on team decision theory, this volume provides a comprehensive study of the general theory of decision making under uncertainty and decentralized information, when the underlying random variables are discrete.

[72-3] **Aoki, M.**, "On feedback stabilizability of decentralized dynamic systems," *Automatica*, 8, 163-173, 1972.

Conditions for stabilizability of decentralized linear time-invariant dynamical systems with partial information are obtained when communication between agents may or may not be permissible. It is shown that controllability does not necessarily imply stabilizability for decentralized systems.

[72-4] **McGuire, C. B.** and **R. Radner**, (eds.), *Decision and Organization*, North Holland, Amsterdam, 1972.

A collection of papers on different aspects of team and organization theory, providing an up-to-date (as of 1972) survey from an economics viewpoint. Papers in this volume place emphasis on information and decentralization.

[73-1] **Groves, T.**, "Incentives in teams," *Econometrica*, 41, 617-631, 1973.

The problem of inducing the members of an organization to behave as if they formed a team is analyzed. A conglomerate type organization consisting of a set of semiautonomous subunits and coordinated by the organization head is considered. The head's incentive problem is to choose a set of employee compensation rules that will induce the subunit managers to communicate accurate information and take optimal decisions. An optimal incentive structure that leads to team behavior is derived, with particular attention being paid to the informational aspects of the problem.

[73-2] **Ho, Y. C.** and **K. C. Chu**, "Equivalence of information structures in static and dynamic team problems," *IEEE Trans. on Automat. Contr.*, AC-18, 187-188, 1973.

For problems with possibly nonlinear dynamics and nonGaussian noises, a sufficient condition for the equivalence of dynamic and static information structures is derived. This may be useful for cases where explicit solutions for static information structures may be easier to find or the existence and uniqueness results for the static version are separately available.

[73-3] **Aoki, M.**, "On decentralized linear stochastic control problems with quadratic cost," *IEEE Trans. on Automat. Contr.*, AC-18, 243-250, 1973.

A decentralized stochastic control problem with agents having different *a priori* information on the initial state is considered, with no observation exchange allowed between the agents. It is shown that the person by person optimal solutions may be found by adding certain correction terms to the solution of the problem under the assumption of centralized information.

[73-4] **Chu, K.**, "Estimation and decision for linear systems with elliptical random processes," *IEEE Trans. on Automat. Contr.*, AC-18, 499-504, 1973.

It is shown that the problem of obtaining team optimal solutions for linear-quadratic systems is tractable when the random variables admit elliptical probability distributions.

[73-5] **Bismut, J.**, "An example of interaction between information and control: The transparency of a game," *IEEE Trans. on Automat. Contr.*, AC-18, 518-522, 1973.

A two person team is studied, where the second agent has perfect information on the action of the first agent, while he does not have access to the measurement based on which the action was selected. It is shown that, in general, there is no optimal control, and that the set of admissible controls can be imbedded densely in the class of controls where the second agent knows the observation of the first agent.

[74-1] **Ho, Y. C. and K. C. Chu**, "Information structures in dynamic multiperson control problems," *Automatica*, 10, 341-351, 1974.

Largely a survey, the paper discusses the static versus the dynamic nature of information structures. For static problems the order of decision making is irrelevant, whereas in the dynamic situation the (noisy) measurements necessarily depend upon some or all of the actions taken at an earlier time. Information may be represented by adopting a "field" point of view, and the concept of nested information comes out naturally as a result of this point of view, as also elucidated earlier in [71-3].

[74-2] **Sandell, N. and M. Athans**, "Solution of some nonclassical LQG stochastic decision problems," *IEEE Trans. on Automat. Contr.*, AC-19, 108-116, 1974.

The linearity of optimal control laws for the one step delay LQG stochastic control problem is established and explicit formulae are presented. Here the one step delay information sharing pattern allows each agent to have access to the information and control of his precedents with a one step delay; that is, it is a special type of a partially nested information pattern. Two related papers are [74-3] and [75-1].

[74-3] **Kurtaran, B. and R. Sivan**, "Linear quadratic Gaussian control with one step delay sharing pattern," *IEEE Trans. on Automat. Contr.*, AC-19, 571-574, 1974.

The optimal controller for discrete time LQG systems with one step delay sharing information pattern is shown to obey a certain "separation" property, and explicit discrete time equations are derived whose solutions are the gains of the optimal controller. Two related papers are [74-2] and [75-1].

[74-4] **Whittle, P. and J. Rudge**, "The optimal linear solution of a symmetric team control problem," *J. Appl. Prob.*, 11, 377-381, 1974.

A symmetric team problem (i.e., one where the control of each agent has identical coefficients in the state equation and cost functional, and the observations of each agent are affected by independent noises having identical variances) is considered, and recursive equations for computing the optimal *linear* controller are found.

[75-1] **Yoshikawa, T.**, "Dynamic programming approach to decentralized control problems," *IEEE Trans. on Automat. Contr.*, AC-20, 796, 1975.

By applying techniques of dynamic programming, problems of decentralized control of stochastic discrete time dynamic systems with one step delay information sharing can be decomposed into several static team problems. This approach gives an optimal control policy for the LQG case, which consists of an optimal control term under the classical information structure and a correction term due to one step delay information structure. Two related papers are [74-2] and [74-3].

[75-2] **Başar, T.**, "Equilibrium solutions in two person quadratic decision problems with static information structures," *IEEE Trans. on Automat. Contr.*, AC-20, 320-328, 1975.

A general class of two person, two objective decision problems characterized by quadratic cost functions and static information structures are considered. Sufficient conditions are derived for the existence of a unique pair of equilibrium solutions assuming that the primitive random variables have arbitrary (but known) probability distributions. For the case of jointly Gaussian statistics, the unique equilibrium solution is shown to be affine in the available decentralized information.

[75-3] **Witsenhausen, H. S.**, "The intrinsic model for discrete stochastic control," *Control Theory, Numerical Methods and Computer Systems Modelling*, in *Lecture Notes in Economics and Mathematics*, 107, 322-335, Springer-Verlag, 1975.

A mathematical model is described within which classical and quasiclassical information structures may be dealt with, bringing some of the problems of nonclassical stochastic control into focus. It is noted for the first time that information theoretic bounds can provide solutions to some nonclassical stochastic control problems.

[75-4] **Ho, Y. C. and G. Hexner**, "Redundancy in team problems," *IEEE Trans. on Automat. Contr.*, AC-20, 439-440, 1975.

This short note introduces the notion of redundancy as an optimally behaving decision maker in a team ignoring a seemingly useful piece of information.

[75-5] **Kurtaran, B.**, "A concise derivation of the LQG one-step-delay sharing problem solution," *IEEE Trans. on Automat. Contr.*, AC-20, 808-810, 1975.

The LQG team problem with the one-step-delay sharing pattern is solved by applying Radner's result [62-1] on static teams to a version of the problem with transformed cost. For related results see [74-2], [74-3] and [75-1].

[76-1] **Kurtaran, B.**, "Decentralized stochastic control with delayed sharing information pattern," *IEEE Trans. on Automat. Contr.*, AC-21, 576-581, 1976.

A two-input, two-output nonlinear and possibly time varying system perturbed by white noise processes is considered, under a single cost criterion, and a general nonclassical separation principle is obtained using dynamic programming. When specialized to linear Gaussian systems, where the optimal controller is known to be linear, the best linear controller is shown to obey a separation principle.

[76-2] **Whittle, P. and J. F. Rudge**, "Simultaneous Communication and Control," *Adv. Appl. Prob.* 8, 365-384, 1976.

Asymptotically optimal policies for dynamic two person teams with nonclassical information are found using the capacity of Gaussian vector channels and the rate distortion function of a vector Gaussian source. The solutions which are, in general, nonlinear in the observations, are unrealizable in control theoretic terms, since the actions depend explicitly on observations lying in the future.

[76-3] **Aumann, R. J.**, "Agreeing to disagree," *Ann. Statistics*, 4, 1236-1239, 1976.

It is shown that in distributed decision making with a common cost functional the decisions of the agents agree if each such decision is the evaluation of the posterior probability of some event and if all agents' posteriors are common knowledge. (This result was later extended in different directions in [82-1], [84-2] and [87-3].)

[76-4] **Başar, T.**, "On the optimality of nonlinear designs in the control of linear decentralized systems," *IEEE Trans. on Automat. Contr.*, AC-21, 797, 1976.

This note provides a counterexample which shows that in the decentralized optimum control of linear deterministic systems with multiple agents and instantaneous output feedback, nonlinear designs could be superior to the best linear ones when the initial state is random.

[76-5] **Rudge, J. F.**, "Series solutions to static team control problems," *Mathematics of Operations Research*, 1, 67-81, 1986.

It is shown that for the static (nonGaussian) team problem with quadratic cost function, there exists a series form for the unique team-optimal solution, which allows the solution to be computed numerically, to any desired degree of accuracy.

[78-1] **Yoshikawa, T.**, "Decomposition of dynamic team decision problems," *IEEE Trans. on Automat. Contr.*, AC-23, 627-632, 1978.

The problem of partitioning a dynamic team problem into several smaller dynamic team problems is studied, and precedence and nestedness relations among groups of team members are defined, leading to the useful concept of independent partition/sequential partition of a team.

[78-2] **Ho, Y. C., M. Kastner and E. G. Wong**, "Teams, signalling and information theory," *IEEE Trans. on Automat. Contr.*, AC-23, 305-312, 1978.

This is an attempt towards unifying results from team decision theory, economic theory and information theory. Some of the problems arising due to dynamic information structure, such as the resulting functional optimization, are made explicit.

[78-3] **Sandell, N. R., P. P. Varaiya, M. Athans and M. G. Safanov**, "Survey of decentralized control methods for large scale systems," *IEEE Trans. on Automat. Contr.*, AC-23, 108-128, 1978.

Largely a survey; the authors note that the determination of optimal strategy for nonclassical stochastic control problems requires, in general, the solution to an infinite dimensional, nonconvex optimal control problem which would also be very difficult to implement. The authors convey the message that it may be unreasonable to seek a single best optimal solution, and therefore one should work with a set of partially ordered and nondominated distributed information and control structures.

[78-4] **Yoshikawa, T. and H. Kobayashi**, "Separation of estimation and control for decentralized stochastic control systems," *Automatica*, **14**, 623-628, 1978.

Using the concept of information structure partitioning, two conditions (one on the independence of random vectors and the other on the form of the cost function) are given, which are sufficient for a separation of estimation and control in decentralized discrete time stochastic systems. While the one step delay sharing pattern satisfies these conditions, a counterexample shows that separation will not necessarily hold for information patterns with more than one step delay.

[78-5] **Başar, T.**, "Decentralized multicriteria optimization of linear stochastic systems," *IEEE Trans. on Automat. Contr.*, **AC-23**, 233-243, 1978.

It is shown that for LQG stochastic teams, the *unique* team optimal solutions can be obtained recursively under the one step delay observation sharing pattern, this result being obtained as a special case of a theorem proved for multicriteria optimization of LQG stochastic systems.

[78-6] **Başar, T.**, "Two-criteria LQG decision problems with one step delay observation sharing pattern," *Information and Control*, **38**, 21-50, 1978.

Person by person optimal solutions,, which are also team optimal, are obtained for a class of LQG teams with decentralized one step observation sharing patterns, by first obtaining results for general two-criteria games and then specializing these to the case when the criteria are identical.

[78-7] **Laub, A. J. and F. N. Bailey**, "An iterative coordination approach to decentralized decision problems," *IEEE Trans. on Automat. Contr.*, **AC-23**, 1031-1036, 1978.

Coordination procedures are examined in an organization-theoretic framework, for the iterative solution of the classical linear least-squares problem in Hilbert space.

[78-8] **Delebecque, F. and J. P. Quadrat**, "Contribution of stochastic control singular perturbation averaging and team theories to an example of large scale systems: Management of hydropower production," *IEEE Trans. on Automat. Contr.*, **AC-23**, 209-221, 1978.

The paper provides a numerical solution to a team problem which arises in the management of a hydropower production system. The team problem is an approximate optimization model which uses local feedbacks.

[78-9] **Schoute, F. C.**, "Symmetric team problems with multi access wire communication," *Automatica*, **14**, 255-269, 1978.

The paper considers a symmetric team problem associated with a large number of communication stations using the same common communication line. The team objective is to use the common channel as efficiently as possible.

[78-10] **Varaiya, P.** and **J. Walrand**, "On delayed sharing patterns," *IEEE Trans. on Automat. Contr.*, **AC-23**, 443-445, 1978.

For delayed sharing patterns Witsenhausen had conjectured that in order to find the optimal control law it may be sufficient to restrict attention to the class of all "separated" laws, (71-1, assertions 8 and 9). The paper provides a counterexample to this conjecture when the delay is two units or more, and proves the conjecture for the unit delay case.

[79-1] **Ho, Y. C.** and **N. Papadoupoulous**, "Further notes on redundancy in teams," *IEEE Trans. on Automat. Contr.*, **AC-24**, 323-325, 1979.

For a class of LQG team problems it is shown that an information structure can be designed so that each decision maker has access to only some part of the overall information, and controls a component of the control vector. Such an information structure is shown to yield the same payoff as the fully centralized one, thus extending the results of [75-4].

[79-2] **Arrow, K. J.** and **R. Radner.**, "Allocation of resources in large teams," *Econometrica*, **47**, 361-385, 1979.

A team with many processes is considered, with payoff depending upon the actions of a resource manager as well as those of a local process manager. A comparison is made between the payoff for the case when the process manager has only local information and the case when he has full information, and it is shown that for "large" teams this additional information has approximately no value if the process parameters are independent and identically distributed and certain "regularity" conditions hold. However, full information is superior when production parameters are correlated.

[79-3] **Başar, T.** and **H. Selbuz**, "Closed-loop Stackelberg strategies with applications in the optimal control of multilevel systems," *IEEE Trans. on Automat. Contr.*, **AC-24**, 166-179, 1979.

For linear-quadratic discrete time systems, controlled by two agents, it is shown that there exists a memory representation of the team optimal strategy for one agent, which preserves its optimality even if the other agent has a different (but known) cost function to minimize (see also [80-6]).

[79-4] **Speyer, J. L.**, "Computation and transmission requirements for a decentralized linear-quadratic-Gaussian control problem," *IEEE Trans. on Automat. Contr.*, **AC-24**, 266-269, 1979.

A decentralized LQG control problem with multiple nodes (each having sensors and controls) is presented, the objective being to optimally share the information of the sensors (processed with a local Kalman filter) with the other nodes. Assuming that the dimensions of the data vector at each node and the state space are large compared with the dimension of the control vector at each node, it is shown that the data to be sent from node to node, to form the optimal controller, can be compressed to a minimal dimension.

[80-1] **Ho, Y. C.** and **T. S. Chang**, "Another look at the nonclassical information structure problem," *IEEE Trans. on Automat. Contr.*, AC-25, 537-540, 1980.

It is established that a partially nested structure is the only one that is equivalent to a static information structure, when the information structures are restricted to be linear.

[80-2] **Bagchi, A.** and **T. Başar**, "Team decision theory for linear continuous time systems," *IEEE Trans. on Automat. Contr.*, AC-25, 1154-1161, 1980.

A team decision theory for linear quadratic continuous time systems is developed, and an iterative convergent scheme is provided, which yields in the limit the optimal team strategies. For the special case of LQG teams, it is shown that the optimal solution is affine. The theory is then extended to dynamic teams with sampled Gaussian measurements and one step delay observation sharing pattern. The unique solution is shown to exhibit certainty equivalence.

[80-3] **Ho, Y. C.**, "Team decision theory and information structures," *Proc. of IEEE*, 68, 644-654, 1980.

This tutorial - survey paper introduces and explains various concepts and results of team theory, as applied to economics, information theory and decentralized control.

[80-4] **Başar, T.**, "A trace minimization problem with applications in joint estimation and control under nonclassical information," *J. of Optimization Theory and Applications*, 31, 343-359, 1980.

The solution of a trace minimization problem under a quadratic inequality constraint is obtained. This solution is then used to obtain the explicit solution to a decentralized stochastic team problem with nonclassical information, which involves joint estimation and control.

[80-5] **Başar, T.**, "Hierarchical decision making under uncertainty," in P. T. Liu (ed.), *Dynamic Optimization and Mathematical Economics*, Plenum Press, 205-222, 1980.

It is shown that Stackelberg solution in two-person stochastic decision problems with two criteria could be team-optimal if the information structure is nested. (See also [84-5].)

[80-6] **Başar, T.** and **G. J. Olsder**, "Team-optimal closed-loop Stackelberg strategies in hierarchical control problems," *Automatica*, 16, 409-414, 1980.

This paper extends the results of [79-3] to continuous-time linear systems controlled by two agents, in which context it is shown that a particular memory representation of the team optimal solution minimizing one agent's cost function remains optimal even if the other agent has a different cost function to minimize. Both finite and infinite horizon cases are considered.

[81-1] **Witsenhausen, H. S.**, "On team guessing with independent information," *Mathematics of Operations Research*, 6, 293-304, 1981.

Exact lower bounds are obtained on the average probability of error a team makes when each agent is confronted with the task of independently detecting an event (the same event for all agents) under independent noisy observation on the occurrence of the event. Here the event is assumed to occur with probability 1/2; a more general case under some other restrictions is considered in [83-2].

[82-1] **Borkar, V. and P. P. Varaiya**, "Asymptotic agreement in distributed decision making," *IEEE Trans. on Automat. Contr.*, AC-26, 650-655, 1982.

A problem in distributed decision making is addressed, where the objective is to arrive at the best estimate of a common random variable through distributed and decentralized communication. Consider the scenario of a stochastic team problem, where at some point in time, some agent receives new information from the environment or a new estimate from some other agent. The paper shows that provided each agent belongs to a ring infinitely often and knows that he is indeed a ring member, then the estimates of all agents asymptotically agree, with the limit depending on the order in which the estimates are transmitted.

[82-2] **Saksena, V. and T. Başar**, "A multimodel approach to stochastic team problems," *Automatica*, 18, 713-720, 1982.

Decentralized team strategies for decision makers using different models of the same large scale system are developed, using multiparameter singular perturbations to capture the multimodel nature of the fast dynamic subsystems interconnected through slow dynamic variables. The theory is applied to obtain a multimodel solution to the linear quadratic dynamic team problem with sampled observations under one step delay observation sharing pattern.

[82-3] **Krainak, J. and S. Marcus**, "Static team problems - Part I: Sufficient conditions and the exponential cost criterion," *IEEE Trans. on Automat. Contr.*, AC-27, 839-848, 1982.

The stationarity conditions of [62-1] are shown, under relaxed hypotheses, to be sufficient to establish the global optimality of candidate decision rules for static teams with convex cost. As an application of this result, global optimality of affine laws for static teams with exponential cost criteria is established, thereby enlarging the class of performance indices for which the optimal solution was known.

[82-4] **Krainak, J. C., J. S. Speyer and S. I. Marcus**, "Static team problems — Part II: Affine control laws,, projections, algorithms, and the LEGT problem," *IEEE Trans. on Automat. Contr.*, AC-27, 848-859, 1982.

Gaussian static team problems with quadratic or exponential cost are solved as constrained parameter optimization problems when the policies are restricted to be affine. In this context a numerical algorithm is proposed.

[82-5] **Krainak, J. C., F. W. Machell, S. I. Marcus and J. L. Speyer**, "The dynamic linear exponential Gaussian team problem," *IEEE Trans. on Automat. Contr.*, AC-27, 860-869, 1982.

For LQG dynamic teams with exponential cost and one step delay sharing pattern, it is shown that the optimum team solution is affine. As an application of this result a resource allocation problem associated with a defense network is considered.

[82-6] **Papadimitriou, C. H. and J. N. Tsitsiklis**, "On the complexity of designing distributed protocols," *Information and Control*, 53, 211-218, 1982.

It is shown that the problem of deciding whether or not two disjoint agents can arrive at compatible decisions without any communication can be solved in polynomial time if there are only two possible decisions for each agent, but it is N-P complete if at least one agent has three or more alternatives.

[82-7] Ho, Y. C., P. B. Luh and G. J. Olsder, "A control-theoretic view on incentives," *Automatica*, 18, 167-179, 1982.

A tutorial survey of literature on "incentives" from a control-theoretic view point. Of particular relevance to this bibliography is the attainment of team-optimal performance for one agent in spite of the fact that some of the other agents may be optimizing different (but known) cost functionals.

[82-8] Başar, T. and J. B. Cruz, Jr., "Concepts and methods in multiperson coordination and control," in *Optimization and Control of Dynamic Operational Research Models*, S. G. Tzafestas (ed.), North Holland, 11, 351-394, 1982.

In this survey paper some key concepts and methods relevant to multiperson decision making and optimization are discussed. The framework is that of dynamic teams and games with decentralized information patterns, which are defined in discrete time.

[83-1] Cansever, D. H. and T. Başar, "A minimum sensitivity approach to incentive design problems," *Large Scale Systems*, 5, 233-244, 1983.

Even though the general result is obtained in the context of nonzero-sum games, it does admit an interpretation in the context of quadratic teams: It is possible to design a "robust" incentive scheme in team problems with a hierarchical decision structure, under which team optimality is preserved even though some of the agents deviate from the overall team goal and the agent at the top of the decision hierarchy does not know the specifics of this deviation.

[83-2] Witsenhausen, H. S., "Team guessing with lacunary information," *Mathematics of Operations Research*, 8, 110-121, 1983.

A sharp bound is obtained on the probability of error agents make when guessing independently on the occurrence of an event, based on decentralized information. (See also [81-1].)

[84-1] Stabile, D. A. and A. H. Levis, "The design of information structures," *Large Scale Systems*, 6, 123-132, 1984.

The paper addresses the problem of designing the allocation of information processing tasks to organization members who interact with the organization's environment. Two information structures are studied, which reduce the load on each agent while accomplishing the overall task.

[84-2] Tsitsiklis, J. N. and M. Athans, "Convergence and asymptotic agreement in distributed decision problems," *IEEE Trans. on Automat. Contr.*, AC-29, 42-50, 1984.

The set-up and main results of [82-1] are extended to convex teams, where estimators are replaced by general decision rules. Results on asymptotic agreement and convergence in the mean square sense (in some cases with probability 1) are obtained for general stochastic teams.

[84-3] Washburn, R. B. and D. Teneketzis, "Asymptotic agreement among communicating decision makers," *Stochastics*, 103-129, 1984.

Results on asymptotic agreement in distributed detection are obtained, where the goal is to reach agreement of a single hypothesis, under decentralized information.

[84-4] **Başar, T.**, "Theory of dynamic games and its applications in large scale systems design and optimization," *Proc. IFAC Congress*, **11.1/A-5**, 177-181, 1984.

The paper discusses the general role of memory strategies in the coordination of interconnected systems when there is a deterministic uncertainty in the modeling and in the quantification of objectives, which may be perceived differently by different decision makers. Even though the general framework is that of dynamic game theory, the general methodology discussed here also applies to team problems where decision makers adopt the same nominal model and objective, but are also allowed to incur small unilateral perturbations around these nominals.

[84-5] **Başar, T.**, "Affine incentive schemes for stochastic systems with dynamic information," *SIAM J. Control and Optimization*, **22**, 199-210, 1984.

It is shown that in two agent stochastic decision problems with a hierarchical structure, there exists a policy for one agent (the leader) under which the optimal team performance is achieved, even though the other agent minimizes a different (but known) strictly convex cost function. The optimal policy for the leader is affine in the dynamic information and generally nonlinear in the static (common or private) information.

[85-1] **Başar, T.**, "An equilibrium theory for multiperson decision making with multiple probabilistic models," *IEEE Trans. on Automat. Contr.*, **AC-30**, 118-132, 1985.

An equilibrium theory for two person quadratic decision problems with static information patterns is developed, wherein the decision makers do not necessarily have the same perception of the underlying probability space. When interpreted in the context of team decision theory, the results of this paper provide valuable insight into the sensitivity and robustness of team-optimal solutions to independent variations in the perceptions of the agents of the underlying probability measure. In particular, in the LQG (linear quadratic Gaussian) set-up, it is shown that the symmetric mode of decision making leads to structurally robust solutions (team solution being extended to Nash equilibrium solution in this case) whereas under the asymmetric mode of decision making the team solution is not robust (here the Stackelberg equilibrium replaces the team solution).

[85-2] **Tsitsiklis, J. N. and M. Athans**, "On the complexity of decentralized decision making and detection problems," *IEEE Trans. on Automat. Contr.*, **AC-30**, 440-446, 1985.

The computational complexity of the discrete versions of some simple but basic decentralized detection problems is studied. The results point to the inherent difficulties in decentralized decision making and suggest that optimality may be an elusive goal.

[85-3] **Başar T. and S. Li**, "Recursive algorithms for optimal solutions of stochastic teams with decentralized information," *Proc. American Control Conf.*, Boston, MA, 231-236, 1985.

The paper extends the framework and some of the results of [85-1] to teams with n(≥ 3) agents, and shows that there are certain aspects of the equilibrium solution which are not shared with the two agent version.

[85-4] **Cansever, D. H. and T. Başar**, "Optimum/near optimum incentive policies for stochastic decision problems involving parametric uncertainty," *Automatica*, **21**, 575-584, 1985.

For a general class of stochastic two-agent incentive decision problems in which the leader (L) has access to the control value of the follower (F) and to private as well as common static information, there exists a smooth control policy for L, which induces the desired behavior on F, even when F's objective function is parametrically uncertain to L.

[85-5] **Cansever, D. H. and T. Başar**, "On stochastic incentive control problems with partial dynamic information," *Systems and Control Letters*, **6**, 69-75, 1985.

This paper extends the main result of [84-5] to decision problems with N>1 followers, where the leader's dynamic information comprises only a linear combination of the followers' actions. It is shown that there exists a policy for the leader, affine in the dynamic information, under which the team optimum value of the leader's objective functional is achieved.

[86-1] **Witsenhausen, H. S.**, "A simple bilinear optimization problem," *Systems and Control Letters*, **8**, 1-4, 1986.

It is shown that the problem of maximizing the inner product of two vectors lying in specified sets of the n-space is N-P complete even when the constraining sets are polytopes; however, an easy solution may be found when both sets are spheres. This result has implications for team theory since it can be shown that the general discrete static team problem is equivalent to the problem of minimizing the inner product of two vectors each prescribed to be in a given set.

[86-2] **Papadimitriou, C. H. and J. N. Tsitsiklis**, "Intractable problems in control theory," *SIAM J. on Control and Optimization*, **24**, 639-654, 1986.

The N-P completeness of the discrete version of Witsenhausen's counterexample [68-1] is established. A theory of continuous algorithms and their complexity is developed, which helps in relating the complexity of the discrete version of a control problem to the complexity of the continuous counterpart.

[86-3] **Teneketzis, D. and N. R. Sandell, Jr.**, "The decentralized disruption problem with linear penalties on false alarms," *Large Scale Systems*, **10**, 227-242, 1986.

This paper addresses team decision theory in the context of decentralized detection. Two detectors take independent noisy observations of a two state Markov chain and have to decide (without communicating with each other) on the time when the Markov chain jumps from one state to another. It is shown that the optimal decision rules are of the threshold types when the cost function couples the decisions of the two detectors and places linear penalty on false alarms as well as on delays in the detection of the jump.

[87-1] **Bansal, R. and T. Başar**, "Stochastic teams with nonclassical information revisited - When is an affine law optimal?", *IEEE Trans. on Automat. Contr.*, **AC-32**, 554-559, 1987.

A parametrized family of two stage LQG stochastic control problems with nonclassical information patterns is considered, and information theoretic bounds are used to show that whenever the performance index does not contain a term involving the product of the decision variables, the optimum solution is linear in the observation variables.

[87-2] **Li, S. and T. Başar**, "Asymptotic agreement and convergence of asynchronous stochastic algorithms," *IEEE Trans. on Automat. Contr.*, **AC-32**, 612-618, 1987.

Results are presented on the convergence and asymptotic agreement of a class of asynchronous stochastic distributed algorithms which are in general time-varying and memory-dependent. Such algorithms arise in both team and game problems, and involve distributed learning and computation.

[87-3] Bansal, R. and T. Başar, "Solutions to a class of linear-quadratic-Gaussian (LQG) stochastic team problems with nonclassical information," *Systems and Control Letters*, 9, 125-130, 1987.

A decentralized two-person stochastic team problem with nonclassical information is studied, with the action of the first agent being transmitted to the second agent through a number of noisy Gaussian channels simultaneously. For the case where all channel noises are independent of the Gaussian input, it is shown that the optimal strategies are linear and may be found via solving a related parameter optimization problem, even when the first agent observes a noisy version of the input.

[88-1] Witsenhausen, H., "Equivalent stochastic control problems," *Math. Control, Signals and Systems*, 1, 3-11, 1988.

A notion of equivalence between stochastic control problems, suitable for complexity analysis, is developed. It is shown that a large class of such problems with dynamic information can be reduced to equivalent static problems, albeit with a transformed cost function, implying that (contrary to common belief, as to be deduced from [80-1]) dynamic decentralized control problems are no harder than static team problems.

[88-2] Tsitsiklis, J. N., "Decentralized detection by a large number of sensors," *Math. Control Signals Systems*, 1, 167-182, 1988.

A decentralized detection problem is considered, with N independent, identical sensors which transmit information to a fusion center which then decides which one of M hypotheses is true. It is shown that as the number of sensors tend to infinity, it is asymptotically optimal to divide the sensors into M(M-1)/2 groups, with all sensors within a group using the same decision rule for transmission. For the case with two hypotheses and a binary message, it is shown that it is asymptotically optimal to have all sensors perform an identical likelihood ratio test.

Lecture Notes in Control and Information Sciences

Edited by M. Thoma and A. Wyner

Vol. 81: Stochastic Optimization
Proceedings of the International Conference, Kiew, 1984
Edited by I. Arkin, A. Shiraev, R. Wets
X, 754 pages, 1986.

Vol. 82: Analysis and Algorithms of Optimization Problems
Edited by K. Malanowski, K. Mizukami
VIII, 240 pages, 1986.

Vol. 83: Analysis and Optimization of Systems
Proceedings of the Seventh International Conference of Analysis and Optimization of Systems
Antiba, June 26-27, 1986
Edited by A. Bensoussan, J. L. Lions
XVI, 901 pages, 1986.

Vol. 84: System Modelling and Optimization
Proceedings of the 12th IFIP Conference
Budapest, Hungary, September 2–6, 1985
Edited by A. Prékopa, J. Szelezsán, B. Strazicky
XII, 1046 pages, 1986.

Vol. 85: Stochastic Processes in Underwater Acoustics
Edited by Charles R. Baker
V, 205 pages, 1986.

Vol. 86: Time Series and Linear Systems
Edited by Sergio Bittanti
XVII, 243 pages, 1986.

Vol. 87: Recent Advances in System Modelling and Optimization
Proceedings of the IFIP-WG 7/1 Working Conference
Santiago, Chile, August 27-31, 1984
Edited by L. Contesse, R. Correa, A. Weintraub
IV, 199 pages, 1987.

Vol. 88: Bruce A. Francis
A Course in H_∞ Control Theory
XI, 156 pages, 1987.

Vol. 88: Bruce A. Francis
A Course in H_∞ Control Theory
X, 150 pages, 1987.
Corrected - 1st printing 1987

Vol. 89: G. K. H. Pang/A. G. J. McFarlane
An Expert System Approach to Computer-Aided Design of Multivariable Systems
XII, 223 pages, 1987.

Vol. 90: Singular Perturbations and Asymptotic Analysis in Control Systems
Edited by P. Kokotovic, A. Bensoussan, G. Blankenship
VI, 419 pages, 1987.

Vol. 91 Stochastic Modelling and Filtering
Proceedings of the IFIP-WG 7/1 Working Conference
Rome, Italy, Decembre 10-14, 1984
Edited by A. Germani
IV, 209 pages, 1987.

Vol. 92: L. T. Grujić, A. A. Martynyuk, M. Ribbens-Pavella
Large-Scale Systems Stability Under Structural and Singular Perturbations
XV, 366 pages, 1987.

Vol. 93: K. Malanowski
Stability of Solutions to Convex Problems of Optimization
IX, 137 pages, 1987.

Vol. 94: H. Krishna
Computational Complexity of Bilinear Forms
Algebraic Coding Theory and Applications to Digital Communication Systems
XVIII, 166 pages, 1987.

Vol. 95: Optimal Control
Proceedings of the Conference on Optimal Control and Variational Calculus
Oberwolfach, West-Germany, June 15-21, 1986
Edited by R. Bulirsch, A. Miele, J. Stoer and K. H. Well
XII, 321 pages, 1987.

Vol. 96: H. J. Engelbert/W. Schmidt
Stochastic Differential Systems
Proceedings of the IFIP-WG 7/1 Working Conference
Eisenach, GDR, April 6-13, 1986
XII, 381 pages, 1987.

Lecture Notes in Control and Information Sciences

Edited by M. Thoma and A. Wyner

Lecture Notes in Control and Information Sciences

Edited by M. Thoma and A. Wyner

Vol. 116: M.D. Mesarovic, Y. Takahara
Abstract Systems Theory
VIII, 439 pages, 1989

Vol. 117: K.J. Hunt
Stochastic Optimal Control Theory
with Application in Self-Tuning Control
X, 308 pages, 1989.

Vol. 118: L. Dai
Singular Control Systems
IX, 332 pages, 1989

Vol. 119: T.S. Başar, P. Bernhard
Differential Games and Applications
VII, 201 pages, 1989

GPSR Compliance

The European Union's (EU) General Product Safety Regulation (GPSR) is a set of rules that requires consumer products to be safe and our obligations to ensure this.

If you have any concerns about our products, you can contact us on ProductSafety@springernature.com

In case Publisher is established outside the EU, the EU authorized representative is:

Springer Nature Customer Service Center GmbH
Europaplatz 3
69115 Heidelberg, Germany

Batch number: 10704763

Printed by Printforce, the Netherlands